Routledge Revivals

The Economics of Water Utilization in the Beet Sugar Industry

Originally published in 1965, this case study of the beet sugar industry undertaken by George O. G. Löf and Allen V. Kneese illustrates the economic importance of water to industry. This study delves into the history and technology of the beet sugar industry to demonstrate the economic impact of the water environment and how water waste can be reduced in other industries. This title will be of interest to students of environmental studies.

The Economics of Water Utilization in the Beet Sugar Industry

George O. G. Löf and Allen V. Kneese

First published in 1968
by Resources for the Future, Inc.

This edition first published in 2016 by Routledge
2 Park Square, Milton Park, Abingdon, Oxon, OX14 4RN
and by Routledge
711 Third Avenue, New York, NY 10017

Routledge is an imprint of the Taylor & Francis Group, an informa business

A Library of Congress record exists under LC control number: 68016166

ISBN 13: 978-1-138-94446-6 (hbk)
ISBN 13: 978-1-315-67184-0 (ebk)

THE ECONOMICS OF WATER UTILIZATION IN THE BEET SUGAR INDUSTRY

by
George O. G. Löf
and
Allen V. Kneese

RESOURCES FOR THE FUTURE, INC.
1755 Massachusetts Avenue, N.W., Washington, D.C. 20036

Distributed by
The Johns Hopkins Press
Baltimore, Maryland 21218

Resources for the Future is a non-profit corporation for research and education in the development, conservation, and use of natural resources. It was established in 1952 with the co-operation of the Ford Foundation and its activities since then have been financed by grants from that Foundation. Part of the work of Resources for the Future is carried out by its resident staff, part supported by grants to universities and other non-profit organizations. Unless otherwise stated, interpretations and conclusions in RFF publications are those of the authors; the organization takes responsibility for the selection of significant subjects for study, the competence of the researchers, and their freedom of inquiry.

This book is one of RFF's studies in water resources and the quality of the environment, directed by Allen V. Kneese. The manuscript was edited by Doris L. Morton and the illustrations were drawn by Federal Graphics.

Director of RFF publications, Henry Jarrett; *editor,* Vera W. Dodds; *associate editor,* Nora E. Roots.

Printed in the United States of America
Library of Congress Catalogue Card Number 68–16166
Price: $4.00

Preface

Water resource planners and those charged with the regulation of waste disposal have usually had only the crudest information, if any at all, about the economic importance of water to industry. In a number of industries—thermal power, pulp and paper, petroleum refining, canning, and beet sugar, among others—water is a very significant input. Also extremely important to these industries is the ability of water courses to receive and assimilate wastes.

This study of the beet sugar industry is one of several projects instituted by Resources for the Future, Inc. to help clarify the economic impact of the water environment upon industry and to analyze the techniques and costs of changing the impact of industry on the water environment.

Two studies have already been published, both in 1965: *The Economic Demand for Irrigated Acreage*, by Vernon W. Ruttan, and *Water Demand for Steam Electric Generation*, by Paul H. Cootner and George O. G. Löf. Other works in progress are studies of water use in the canning industry, in petroleum refining, and in pulp and paper manufacture.

The authors of the present study wish to acknowledge the co-operation of the beet sugar industry in supplying data. Lloyd Jensen, Vice President of The Great Western Sugar Company, President of the American Society of Sugar Beet Technologists, and former Chairman of the National Technical Task Committee on Industrial Wastes, was especially helpful both in supplying information and in reviewing drafts of the manuscript. The manuscript was also reviewed in detail by Guy Rorabaugh, Vice President of Holly Sugar Corporation, and Hugh G. Rounds, Assistant General Superintendent of The Amalgamated Sugar Company. We also wish to express our gratitude to the officials of the other beet sugar companies in the United States who provided detailed information on their use of water. These companies are the American Crystal Sugar Company, Buckeye Sugars, Inc., Michigan Sugar Company, Monitor Sugar Division of Robert Gage Coal Company, The National Sugar Manufacturing Company, Northern Ohio Sugar Company, Spreckles Sugar Company, Union Sugar Division of Consolidated Foods Corporation, and Utah-Idaho Sugar Company.

Blair T. Bower, of the staff of Resources for the Future, reviewed several versions of the manuscript and made many useful suggestions for its improvement. Others who reviewed the manuscript were Hayse H. Black, Industrial Water Consultant, of Cincinnati; Michael F. Brewer of Resources for the Future; Jack Carlson of the staff of the President's Council of Economic Advisers; Paul H. Cootner of the Massachusetts Institute of Technology; Henry De Graff of the U.S. Department of Commerce; Wesley Eckenfelder of the University of Texas; and Ralph Porges of the Delaware River Basin Commission staff.

July 1967

GEORGE O. G. LÖF
ALLEN V. KNEESE

Contents

Preface v

I. Introduction 1
Concepts of Water Use 1
Some Definitions 5
Plan of the Book 7

II. The Industry and Its Use of Water 9
Some General Characteristics 9
Published Reports on Water Use 12

III. The Water Technology of the Industry 16
The Manufacturing Process 16
Water and Waste Aspects of Production 21
Fluming and washing 21
Water flow through diffusers 23
Further uses of water in production 26
The Steffens process and its waste 29
Implications of Process Changes for Saving Water and Reducing Waste Loads 32
Other Methods for Reducing Waste Loads 34
Mechanical separation of suspended matter 34
Treatment of waste water in plant using biological processes of waste degradation 35
Temporary storage of wastes with either continuous or programmed discharge 36
Storage of all wastes, plus evaporation and seepage 36
Spreading of waste waters over land 36
Industry Practice in Handling Waste Loads 37

IV. Water and Waste in U.S. Plants 39
The Questionnaire 39
Water Withdrawal or Intake 40
Gross Water Use 45
Recirculation 46
Water Sources and Treatment Processes 47
Depletion or Consumptive Use of Water 49

Waste Waters 52
General aspects 52
Waste quantities 52
Waste reduction through recirculation 60
Waste reduction per ton of beets 62
Current Practice and Trends in Water Recirculation and Waste Treatment Processes 62
Continuous diffusers 64
Pulp drying 65
Steffens waste processing 66
Recirculation of condenser water 68
Condenser water reuse for flumes and beet washers 70
Recirculation of flume water and beet-washer water 71
Simultaneous recirculation of flume-washer and condenser streams 73
Waste treatment practices 74
New waste treatment processes 79
Present Costs of Treating Water and Wastes 82

V. Economic Implications 86
General Conclusions 86
Handling Lime Cake Slurry 88
Treating Condenser Water 91
Treating Flume and Beet Washing Water 96
Cost of BOD Reduction 100
Concluding Comments 104

Appendix A. Water Usage Relationships by Regression Analysis, by Robert M. Steinberg and Betty Duenckel .. 107

Appendix B. Questionnaire on Industrial Plant Water Supply and Waste Disposal 112

Appendix C. Water Use Survey Supplementary Questionnaire 124

LIST OF TABLES

1. Sugar Industry Statistics, 1964 10
2. Gross Water Use in Typical U.S. Beet Sugar Plant 20
3. Wastes from Each Process Step in Beet Sugar Manufacture 22
4a. Effects of Various Beet Sugar Plant Features on Trends in Water Usage, as Determined by Regression Analysis of Questionnaire Data 43
4b. Effects of Various Beet Sugar Plant Features on Trends in Water Usage, as Determined from General Characteristics of Processes 44
5. Representative BOD Quantities in Wastes from Plants Having Various Processes and Water Handling Procedures .. 55
6. BOD in Beet Sugar Plant Wastes, 1949 and 1962 58
7. Recirculation and Waste Treatment Processes in Beet Sugar Plants 63

LIST OF FIGURES

1. Schematic Water Flow Diagram 6
2. Raw Materials and Products in the Beet Sugar Industry 17
3. Materials Flow in Beet Sugar Plant with No Recirculation or Treatment of Waste Waters 19
4. Materials Flow in Beet Sugar Plant with "Typical" Water Utilization and Waste Water Disposal Pattern 24
5. Materials Flow in Beet Sugar Plant with Full Recirculation and No External Discharge of Waste Loads 33
6. Cost of Reducing BOD Content of Lime, Flume, and Condenser Water Wastes from 2,700-Ton-a-Day Beet Sugar Plant 102
7. Incremental Cost of Reducing BOD Content of Lime, Flume, and Condenser Water Wastes from 2,700-Ton-a-Day Beet Sugar Plant 103
8. Reduction in BOD Discharged in Lime Slurry, Flume Water, and Condenser Water from 2,700-Ton-a-Day Beet Sugar Plant 105

1

Introduction

Concepts of Water Use

Industry's use of water is highly complex. Taking the beet sugar industry as an example and assuming the need for efficient use of national water resources by all industries, we begin by conveniently dividing the industrial use of water into three categories.

First is *withdrawal.* Industry takes huge amounts of water from water sources. In the United States, manufacturing industries withdraw about twice as much water as municipalities. Of this industrial part, a very small proportion (about .03 per cent) is withdrawn by beet sugar plants. However, because of the high degree of seasonality of operation[1] and considerable geographical concentration, the withdrawal of water by this industry becomes a significant portion of total withdrawals in several regions of the country.

These large withdrawals of water by industry are possible at some locations only if regional storage reservoirs release sufficient streamflow. Also, the quality of the water available for withdrawal may significantly affect the costs of the withdrawing industry. The industry may benefit, for example, by increased dilution water or higher levels of waste treatment upstream. On the other hand, it is possible in most instances for individual industrial firms to reduce withdrawals—by recirculating their water—and it is always possible to improve the quality of intake water by treatment in the plant. (We will see that these are genuine and often feasible alternatives in the beet sugar industry.)

Decisions—both public and private—to regulate the quality and quantity of water sources must, in part, therefore, turn on the technological opportunities for, and the cost of, in-plant re-

[1] About 60–120 days during the fall and winter, except in California where the season tends to be considerably longer.

circulation. And similarly, they must consider the opportunities for quality adjustment and/or costs to the industry of having available only poor quality water. The possibilities of such adjustments make the usual projections of water quantity and quality "requirements" for specific industries questionable, to say the least. One objective of this study is to show the economic effects on the beet sugar industry associated with alternative policies for handling water quality and quantity—both in-plant and region-wide.

A second category of water use by industry, with its accompanying impact on regional water supplies, occurs because of actual *depletion* or *loss* associated with use. In some industries, significant water losses occur due to evaporation in process use or embodiment in products. In the case of beet sugar, actual losses of this kind during beet processing are virtually nonexistent. There is some evaporation in the plant itself, especially where the extracted beet pulp is dried, but this vapor loss is usually made up, or more than made up, by extraction of water from the beets (by draining and pressing of wet spent pulp). It is in the ponds all beet sugar plants have for the temporary storage of certain effluents that loss occurs. Some factories employ ponds large enough to eliminate all waste waters by evaporation and underground seepage. Depletion or evaporation loss may therefore be significant, and when the ponds are used for temporary storage of wastes during low flow, with periods for release during periods of high streamflows, water is withheld from the regional stream system at the most critical times. This study provides information on the water losses in the beet sugar industry associated with various methods of using water and disposing of waste water.

Using water courses for *disposal of industrial wastes* creates the final and usually most important impact on regional water supplies. Indeed, the primary reason for devoting special and detailed attention to the use of water by the beet sugar industry is the highly pollutional character of the wastes generated by certain processes used in beet sugar factories. The statement is often encountered that a single sugar plant may generate wastes with a biochemical oxygen demand (BOD) equal to that of a city of a quarter million or more people. If such a quantity of waste is discharged directly to the small streams which characterize most of the sugar beet producing areas, it will place a serious burden upon dissolved oxygen (DO). Indeed, DO may be depleted to the point where fish are killed or even aesthetically repellant septic conditions produced. Beet sugar factory wastes also carry a large quantity of suspended solids[2] which, unless removed, may form sludge banks in streams. Their decomposition increases the demand on available oxygen and creates objectionable odors and unsightly surroundings.

Furthermore, while most of the wastes produced—we shall focus upon BOD as the single most important—are in principle quite amenable to successful treatment by conventional biological means, these procedures are not usually practical. A trickling filter or an activated sludge plant takes perhaps a month to become fully operative, that is, for the bacterial cultures to develop, whereas the annual production run of the sugar plant is usually only 60–120 days. Thus, in practice, the receiving waters are given only very partial protection. Moreover, the capital investment in such plants is rather heavy, especially if the entire waste water flow from a plant is to be treated and the capital cost amortized over very short periods of operation. Recirculation of various waste-water streams and storage of the remainder in ponds to provide a more even feed over time to the waste treatment plants may improve the economics of treatment,[3] but even so, treatment appears to be less favorable than several other alternatives for reducing waste loads.

The huge BOD loads that can be generated in the processing of sugar beets and the difficulties of successful and economical treatment have been the reasons for stressing beet sugar in connection with the "pollution problem." One influential study that has projected waste loads and in which beet sugar wastes played a significant role is that of the Senate Select Committee on National Water Resources. The analysts producing this

[2] In part, the BOD is contained in the suspended solids.

[3] This technique is practiced at several locations in Germany.

study began with some estimates Rudolfs published in 1953 on the generation of industrial wastes.[4] They then assumed that the relationship between production and waste loads would not change during the period of analysis, which extended to the year 2000. If one assumes that sugar production will increase at the same rate as the broader category "food products," for which an explicit projection was made, an amazing and unreasonable result is obtained. *It is that the waste load coming from this one small part of the food industry will be almost two-thirds as large as the residual load coming from all municipalities in the entire United States at the present time.*

Fortunately, the prospect is not as spectacularly grim as this. As it turns out, beet sugar production processes present really striking opportunities for waste reduction by means of recirculation, process changes, waste recovery and other waste-treatment processes. It is perfectly feasible to design a beet sugar plant which generates no waste requiring external discharge. This can be done with current technology, and, indeed, all the individual devices needed are already in operation at one site or another, here and abroad.

While some of these existing devices were installed because they proved internally profitable to the plant, it is unfortunately not the case that it is usually to the benefit of a given plant to reduce its wastes to a very low level solely because of the *internal* savings which may thereby accrue. Waste reduction usually comes at some cost. But so does waste discharge, albeit not usually to the discharger himself. Discharge often imposes an external or spillover cost on subsequent users of the stream and on the general public. This external cost may be high enough to justify society requiring or inducing industrial plants to reduce their waste loads. An optimum, in the sense of minimizing overall costs—internal and external to the industrial

[4] See *Water Supply and Demand*, Select Committee on National Water Resources, United States Senate, Committee Print No. 32 (Washington: U.S. Government Printing Office, August, 1960), and William Rudolfs, "The Problem," in William Rudolfs, editor, *Industrial Wastes and Their Treatment* (Valley Stream, New York: Library of Engineering Classics, 1953). The data on the beet sugar industry published by Rudolfs relate to 1949.

plant—is reached when waste reduction is carried to the point where the internal cost of the last increment of reduction is just equal to the external cost avoided.

The measurement of external costs is an extremely difficult task which we do not undertake in this study except as such costs might be imposed on the beet sugar industry itself by other waste dischargers.[5] We do, however, hope to provide a relatively firm basis for determining the internal cost which beet sugar plants incur when they cut back their waste loads by various amounts. This is a highly significant datum for any control effort which aims, even roughly, at making optimum use of water resources, including their waste assimilative capacity. It is significant too for efforts aimed at the more limited objective of achieving a specified stream water quality with a minimum cost combination of alternative measures, including process change, increased dilution, waste treatment, measures to improve receiving water assimilative capacity, and so on.[6]

Since procedures to reduce industrial waste discharge so frequently involve recirculation, they also have a strong impact on plant water intake and may be advantageous in areas where water is quantitatively in short supply. Indeed, in some instances the primary motivation for recirculation has been the high cost of obtaining additional water supplies. This suggests that demand for withdrawal responds rather elastically—at least given enough time for adjustment—to the price or cost of water. Exploring the economics of water withdrawal, water losses, and waste water control and their interdependence is the objective of this study.

Some Definitions

In the hope that it will help the reader, we now provide brief definitions of the main concepts relative to water utilization as used in this study. These definitions are illustrated in Figure 1.

[5] Studies of costs imposed on municipal, other industrial (including pulp and paper, canning, thermal power, and petroleum refining), and recreational uses of streams are being conducted under RFF sponsorship.

[6] These concepts are discussed further in Allen V. Kneese, *The Economics of Regional Water Quality Management* (Baltimore: The Johns Hopkins Press for Resources for the Future, Inc., 1964).

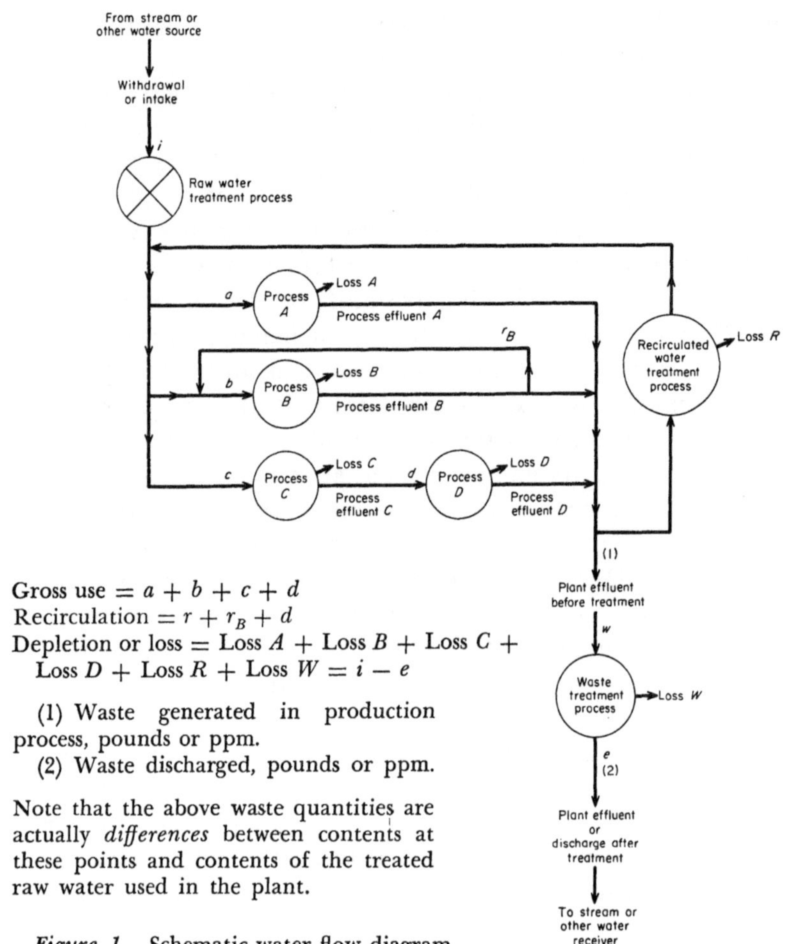

Gross use $= a + b + c + d$
Recirculation $= r + r_B + d$
Depletion or loss = Loss A + Loss B + Loss C + Loss D + Loss R + Loss $W = i - e$

(1) Waste generated in production process, pounds or ppm.
(2) Waste discharged, pounds or ppm.

Note that the above waste quantities are actually *differences* between contents at these points and contents of the treated raw water used in the plant.

Figure 1. Schematic water flow diagram.

Water withdrawal or intake: The volume of water removed from a surface or underground water source (stream, lake, or aquifer) by plant facilities or obtained from some source external to the plant.

Process effluent or discharge: The volume of water emerging from a particular use in the plant.

Water recirculation or recycling: The volume of water already used for some purpose in the plant which is returned with or without treatment to be used again in the same or another process. It is equal to the sum of process effluents minus the sum of plant discharge before treatment and any water which may be lost in the recirculation process itself (evaporation from cooling towers, for example).

Water use or gross use: The total volume of water applied to various uses in the plant. It is the sum of water recirculation and water withdrawal.

Depletion or loss: The volume of water which is evaporated, embodied in product, or otherwise disposed of in such a way that it is no longer available for reuse in the plant or available for reuse by others outside the plant.

Plant effluent or discharge before treatment: The sum of process effluents minus water entering all recirculation systems and successive use processes.

Plant effluent or discharge after treatment: The volume of waste water discharged from the industrial plant. In this definition, any waste treatment device (pond, trickling filter, etc.) is considered part of the industrial plant. The plant effluent equals withdrawal minus depletion.

Waste generated: The amount (usually expressed as weight) of some residual substance generated by a plant process or the plant as a whole and which is suspended or dissolved in water. This quantity is measured before treatment.

Waste discharged: The amount (usually expressed as weight) of some residual substance which is suspecded or dissolved in the plant effluent after treatment, if any.

Plan of the Book

Before proceeding with our actual analysis of the economics of water use and waste disposal in the beet sugar industry, we provide some background on the history and technology of the industry.

Accordingly, in the next chapter we briefly describe the development and location of the industry—briefly, because there are good specialized sources available on this topic. This chapter also reviews some of the information presented in various published sources on water use and waste disposal and makes some comparison with practices in Western Europe.

Following that is a background chapter on technology of the

industry. This is merely meant to set the stage for our statistical and engineering analyses and is in no sense meant to rival the specialized works on the subject.

Thereafter, we turn to presentation and analysis of results of a detailed questionnaire to which all member firms of the industry responded. Here, we have separate sections on water withdrawals, water losses, and disposal of wastes. Because of the importance of the waste question, we present rather detailed discussion of the major, individual (actual or potential) sources of waste loads in the beet sugar manufacturing process.

In the final chapter, we endeavor to show how "typical" plants would optimally reduce waste loads to various specified amounts and what *net* internal costs would be associated with this.

Appendix A presents a regression analysis of data from beet sugar plants, by Robert M. Steinberg and Betty Duenckel. In Appendix B and Appendix C, questionnaires sent to the industry are reproduced.

Even though most of our data are relatively recent (1962), we are aware that further changes have occurred in the industry since then. It has not been possible for us to take explicit, quantitative account of them. We can say, however, that they are consistent with the directions pointed in this book, that is, increasing recirculation of water and reduced waste loads.

2

The Industry and Its Use of Water

Some General Characteristics

About 40 per cent of the nation's sugar supply is produced from sugar beets and cane grown in the continental United States and an additional 12 per cent from cane grown in Hawaii. Another 9 per cent is cane sugar from Puerto Rico and the Virgin Islands, classed under the U.S. Sugar Act as a "domestic area." The rest is imported. The six million tons of sugar currently imported into the continental United States from domestic insular and foreign sources constitute something over one-third of all the sugar moving in world trade. World consumption of sugar is roughly sixty-five million tons, of which the United States accounts for about ten million. Of the total world demand for sugar and syrups, 85 to 90 per cent has been met by cane and beet sugar.[1] Of world production, about 56 per cent comes from cane and about 44 per cent from beets. Table 1 contains some of the more important sugar industry statistics for 1964 and 1965.

Approximately 70 per cent of the sugar consumed in the United States originates from cane. However, only about 10 per cent of this comes from cane grown in the continental United States—in Louisiana and Florida. Sugar other than that produced from cane grown in these two states comes into the continental United States as raw sugar of about 96 per cent purity. The further refining of the sugar does not require large water supplies and the resulting waste discharge is usually small. Consequently, while domestically produced beet sugar accounts for a relatively small proportion (25 to 30 per cent) of domestic sugar consumption, it gives rise to most of the water use and waste disposal problems associated with the sugar industry in the continental United States.

[1] See Hans H. Landsberg, Leonard L. Fischman, and Joseph L. Fisher, *Resources in America's Future* (Baltimore: The Johns Hopkins Press for Resources for the Future, Inc., 1963), p. 103.

TABLE 1
Sugar Industry Statistics[a]

(Millions of short tons unless otherwise indicated)

	Raw sugar		Refined sugar[b]	
	1964	*1965*	*1964*	*1965*
World production cane sugar	36.3	41.0	35.0	38.4
World production beet sugar	29.4	28.9	28.2	27.8
Total world sugar production	65.7	69.9	63.2	66.2
U.S. production cane sugar	3.3	3.2	3.1[c]	3.0[c]
Continental U.S.	1.1	1.1	1.0	1.0
Hawaii	1.2	1.2	1.1	1.1
Puerto Rico and Virgin Is.	1.0	0.9	0.9	0.8
U.S. production beet sugar	3.3	2.9	3.2	2.8
Total U.S. sugar production	6.6	6.1	6.3	5.8
Sugar imports to U.S. from foreign sources (all cane)	3.6	4.0	3.4	3.7
U.S. consumption cane and beet sugar[d]	9.8	10.0	9.3	9.5

	1964	*1965*
U.S. sugar beet acreage harvested (*millions of acres*)	1.40	1.25
U.S. average beet tons per acre	16.80	16.70
U.S. sugar beet production (*millions of short tons*)	23.40	20.90
U.S. sugar beet crop value (*$/ton*)	14.01	13.57
U.S. sugar beet crop value, total (*millions $*)	328.00	284.00

Typical sugar content of beets....................15%
Typical sugar recovery, non-Steffens plant...........85% = 0.13 ton/ton beets
Typical sugar recovery, Steffens plant...............95% = 0.14 ton/ton beets
Typical dried pulp production....................0.05 ton/ton beets
Typical molasses production, non-Steffens plant.......0.045 ton/ton beets
Typical values of products at plant:
Sugar....................................$150 per ton
Dry pulp....................................$45 per ton
Molasses....................................$35 per ton

[a] Raw sugar, beet acreage, beet production, and crop value statistics for 1964 and 1965 taken from U.S. Department of Agriculture, *Agricultural Statistics, 1965* and *Agricultural Statistics, 1966* (Washington: U.S. Government Printing Office), tables 117, 122, 126, 129, and 130 of both volumes.

[b] Refined cane sugar tonnage computed from raw cane sugar figures by assuming average recovery of 93.5 per cent refined white sugar from raw sugar in cane refineries. Refined beet sugar tonnage computed from reported "beet sugar raw values" by employing 96 per cent conversion factor used in determining raw values from actual refined beet sugar production.

[c] Totals slightly different from sums of individual figures, due to rounding.

[d] Consumption may differ slightly from production plus imports, due mainly to inventory changes.

In Western Europe, however, most of the sugar demand is met from beet production, and while the world's largest beet sugar industry is in the U.S.S.R.,[2] the countries of France, Germany, Poland, Italy, Czechoslovakia, and the United Kingdom have large beet sugar outputs.[3] The German industry is of roughly equal size to that in the United States, and England has about one-half the U.S. capacity. Special interest attaches to industries in these areas, because they have operated under conditions of restricted water supply—especially to receive wastes—for many years and consequently have pioneered in the development of processes and devices which will be of considerable interest to us later.

The first beet sugar factory in Europe was put into operation in Breslau, Germany, in 1801. The first successful one in the United States was established at Alvarado, California, in 1879.[4] For a time, the industry developed rapidly. Many factories were built in the first two decades of this century. At the present time, the United States has about sixty sugar factories. Most of the plants themselves are quite old although some of the equipment they use is new.

Factories are located in fifteen states. California and Colorado each process about 20 per cent of the total United States beet crop, Idaho and Minnesota about 10 per cent each, and Oregon, Nebraska, Montana, Michigan, and Washington each process between 4 and 7 per cent of the total. Other states in the same geographical areas as those indicated handle the remainder. In 1964, sugar beets were grown in 20 states.

Generally, the capacity of these plants permits them to process from about 1,000 to 7,000 tons of beets per day. During their seasonal period of operation—in the fall and early winter in

[2] In 1960 the U.S.S.R. produced 7.5 million tons of sugar in 250 factories. From an unpublished paper by S. L. Force, "Beet Sugar Factory Wastes and Their Treatment—Primarily by the Findlay System."

[3] For discussions of the German and English industries, see Friedrich Sierp, *Gewerbliche und Industrielle Abwässer* (Berlin: Springer Verlag, 1959), pp. 260 ff., and T. W. Brandon, "Waste Waters from Beet Sugar Factories," *The International Sugar Journal,* Vol. XLIX, No. 580, pp. 98–100.

[4] Erman A. Pearson and Clair N. Sawyer, "Beet Sugar Process Waters—Treatment and Utilization," *Chemical Engineering Progress*, August, 1950.

most states—they run continuously day and night, 7 days a week for 60–120 days, with a typical production duration of 100 days. This period is known as the "campaign" and is somewhat longer than the beet harvesting period which runs 60–90 days.[5]

Published Reports on Water Use

In almost all considerations of water use in beet sugar manufacture, quantities are most conveniently based on beets processed rather than sugar produced. The most common figures are gallons of water per ton of beets handled. Also dealt with are tons of beets handled per day and gallons of water withdrawn, used (gross), and recirculated per day. Selection of the beet basis, rather than the sugar basis, is convenient because of the industry practice of rating the capacity of a plant in tons of beets processed per day. Moreover, water usage (in a particular plant) is usually closely proportional to beets handled regardless of the high or low sugar content of the beets or the extent of extraction, so the water-to-beet ratio is more nearly constant than the water-to-sugar ratio. On the average, however, the water use figures may be converted to an approximate sugar product basis by using a typical yield of 0.13 ton of sugar per ton of beets. Thus, if there are 5,000 gallons of water used per ton of beets, there would be 38,500 gallons of water per ton of sugar or about 19 gallons per pound of sugar.

Reports on the amount of water used in processing a ton of beets vary widely. To a major extent, the divergence reflects differences in the processes used. But it is also sometimes unclear which concept of use is being employed. Sierp[6] gives a range of about 2,600 to 7,800 gallons per ton as characteristic of Germany. Since he appears to be focusing on withdrawals, this range reflects recirculation patterns. The high figure presumably refers to a plant which practices little or no recirculation. The lower figure, however, appears to be too high for a reasonable minimum withdrawal in view of the results of the German cen-

[5] As noted, both the period of harvesting and the "campaign" are usually longer in California.

[6] Friedrich Sierp, *op. cit.*, p. 260.

sus reported below and our knowledge of the technology of the industry. Brandon[7] suggests a figure of 5,000 gallons of waste water per ton in England, which should about correspond with withdrawal. If this is meant to represent an average, it is certainly very high relative to other estimates of water withdrawal. Phipps[8] cites the *withdrawal* of 3,500 gallons per ton in a British plant in which no water is reused, and an average withdrawal (for 18 factories) at 1,200 gallons per ton accompanied by recirculation of 1,200 gallons, for a gross average use of 2,400 gallons per ton. Porges and Hopkins[9] of the United States indicate that 2,500 to 3,500 gallons of water are used per ton. While the authors say "use," they presumably mean "gross use minus recirculation," or what we have defined as withdrawal. As will be indicated later, many beet sugar plants now withdraw far less water than indicated in the reports cited above. If withdrawals are as low as a very few hundred gallons per ton, vapor losses may become a significant fraction of withdrawal; waste water volume may be appreciably reduced thereby.

Once more we draw attention to the importance of making a careful distinction between several concepts of use[10] and we review briefly for the reader as follows. *Withdrawal* or *intake* refers to the amount of water diverted or pumped from a source for use in the plant. *Gross use* refers to this same amount plus the quantity which is supplied for reuse by recirculation. *Depletion* indicates the amount of water which the plant actually loses, mainly by evaporation. *Discharge* refers to the amount of water which leaves the plant and enters a water course or a holding lagoon from which there is no surface flow. In present-

[7] T. W. Brandon, p. 98.

[8] O. H. Phipps, "National Reports, Great Britain," in J. M. Henry, "The Problems of Waste Waters in Sugar Factories," in International Union of Pure and Applied Chemistry, *Re-Use of Water in Industry* (London: Butterworths & Co., 1963), pp. 233–37.

[9] Ralph Porges and Glen J. Hopkins, "Broad Field Disposal of Beet Sugar Wastes," in *Sewage and Industrial Wastes,* Vol. 27, No. 10 (October 1955), p. 1169.

[10] See Blair T. Bower, "The Economics of Industrial Water Utilization," in Allen V. Kneese and Stephen C. Smith, editors, *Water Research* (Baltimore: The Johns Hopkins Press for Resources for the Future, Inc., 1966), pp. 144–49.

ing our own data in later chapters, we again draw a distinction between plant effluent or discharge before and after treatment. Treatment may occur in various facilities including lagoons.

Fortunately, information supplied by both the United States and West German industrial censuses can be made to fit the above definitions. Censuses incorporating information on water use in beet sugar factories are available for the years 1954 and 1959 for the United States and 1955 and 1957 for Germany. On the whole, the later censuses of both countries are more complete and appear to be more reliable than the earlier ones. Since even the later censuses are limited in accuracy, the numbers presented below have been rounded drastically.[11]

	U.S.	*Germany*
Intake	2,600	1,400
Gross use	4,400	5,000
Discharge total	2,400	1,300
Treated in plant	900	560
Untreated	1,500	750
Untreated discharge to public sewer	negligible	270
Untreated discharge to surface water, ground or wells	1,500	480

For what they are worth, and we will postpone further assessment until we present our own data, these figures confirm reasonable hypotheses one might hold about water use in the two countries. While the German industry appears to be slightly "wetter" in the sense of a higher gross use per ton of beets, its actual withdrawal from water sources is only a little more than half as great as that of the United States. This, of course, reflects significantly more extensive internal recirculation. Accordingly, the volume of waste water is also much lower. Although this does not mean that a BOD tonnage decrease must

[11] The authors are indebted to Vera Eliasberg for collecting and analyzing the German and U.S. Census data. Original sources are: U.S. Bureau of the Census, *Census of Manufactures: 1958*, "Industrial Water Use, 1959," Bulletin MC58 (1)-11 (Washington: U.S. Government Printing Office, 1961); and "Die Wasserversorgung der Industrie im Bundesgebiet, 1957" (Bad Godesberg: Bunderministerium fuer Atomkernenergie und Wasserwirtschaft, 1960).

accompany the reduced waste flow, several of the available recirculation processes also significantly reduce the waste discharged. Unfortunately, neither of the censuses yields information on the constituents or strength of the waste. A slightly higher proportion of the waste discharged is treated by the U.S. plants but, as will be clearer when technology is discussed, it is extremely difficult to infer what this may mean.

An interesting difference is that whereas U.S. plants discharge only negligible amounts of water to public sewer systems, a substantial proportion of the German waste water goes into them. This reflects the more dense industrial and population development surrounding the German beet sugar plants. Presumably, the wastes discharged to public systems receive some further treatment.

One may conclude that at the time of the above census estimates, greater recirculation and higher treatment in Germany resulted in a lower level of actual waste discharge to receiving waters. As shown in a subsequent section of this document, however, major changes in the U.S. beet sugar industry during the last five years or so have greatly reduced or possibly eliminated this difference.

A number of estimates appear in the literature showing the actual amount of BOD (variously measured in pounds or population equivalents) of potential wastes generated by the various processes used. It will be more convenient to discuss these estimates in conjunction with our outline of the technology of the industry.

3

The Water Technology of the Industry

The technology of the beet sugar industry, while not simple, is highly standardized from one plant to the next. All factories in this country and abroad use fundamentally the same processes. Thus, the basic outline of the processes presented here has universal validity. Major differences affecting waste loads, proportion of sugar recovered from the beet, and other variables result from differences in recirculation and waste recovery practices and from facilities and processes which are essentially extensions of the basic plant. This chapter will show how the basic production process, differences in recirculation and waste recovery patterns, and the character of the tertiary processes fundamentally affect water withdrawals, consumptive use, and the quantity of wastes discharged from the plant.

The Manufacturing Process

Before examining the water and waste aspects of the beet sugar industry, we shall take a brief overall view of the sugar production process. This view may be improved if the reader refers to Figures 2 and 4, dealing respectively with the main raw materials and products and with the flow through the principal processing steps.

Sugar beets, limestone, fuel, and a small amount of sulfur are the raw materials of the industry, water is a processing aid or agent, and sugar, beet pulp (wet or dry), and molasses are the main products. Figure 2 shows typical proportions of these materials.

Although the general process of manufacture is the same in all beet sugar plants, some techniques, particularly with respect to water use, are highly variable. The yield of the main product—sugar—is always maximized within economic limitations,

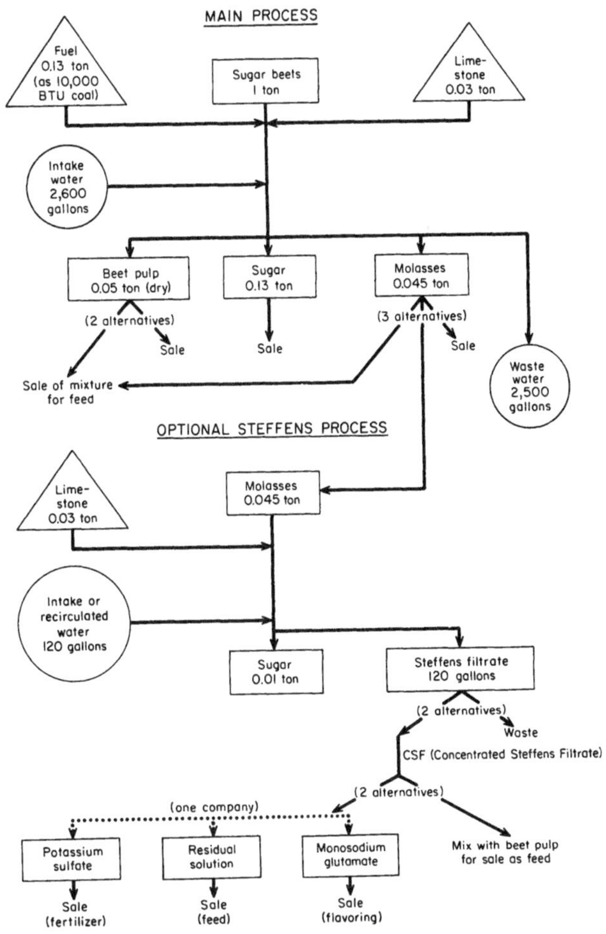

Figure 2. Raw materials and products in the beet sugar industry (typical quantities shown).

but the quantities and characteristics of the pulp and molasses by-products vary from plant to plant and year to year. Beet pulp may be sold wet for immediate stock feeding or dry for feed use as needed. Molasses may be sold directly, for stock feed, alcohol manufacture and yeast production; it may be mixed with beet pulp prior to drying and sold as a higher value feed; or, depending on market conditions and the availability of plant, it may be used in the Steffens process for production of about 10 per cent more sugar. If the Steffens process is used, the low-value Steffens filtrate, if not discarded, is concentrated and mixed with pulp. In one plant, it is used as raw material in the recovery of monosodium glutamate (flavor intensifier for human use), potassium sulfate (for fertilizer), and a liquid concentrate of high protein content for mixing with dried stock feeds.

In addition to this variability in by-product quantities and types, there are variations in sugar content of the beets (a) in different parts of the country due to climatic and soil factors, and (b) in different years due to climatic and other environmental conditions. There are also plant-to-plant differences in the amount of sugar recovered from the beets, mainly because of the use or non-use of Steffens processes, but also because of design and operational factors in the main production line. Sugar recovery percentage is also affected by beet quality, which is, in turn, dependent on the extent of spoilage. Chief factors in spoilage are freezing and subsequent deterioration in storage.

Finally, the greatest variation from plant to plant is in intake water quantity, caused largely by differences in water recirculation methods, and to a lesser extent by variations in design and operation of the sugar production process, climatic differences (affecting condenser water use), and seasonal weather factors (affecting beet cleanliness and wash water use).

In our general view of beet sugar manufacture (see Figure 3), we see that beets are first transported to the plant from storage in a water flume, washed, and then sliced into small noodle-shaped strips. Sugar is dissolved from these slices by use of hot water, and the resulting thin juice is clarified by successive additions of lime and carbon dioxide. After filtration, the juice is concentrated to a thick juice or syrup and finally evaporated (under vacuum) to form a suspension of sugar crystals in a heavy syrup. The crystals are separated from the syrup in a centrifuge, then dried and stored.

The heavy syrup still contains much sugar, so it is recycled for further evaporation and crystallization. The final liquid residue, from which no more sugar can be economically crystallized, is by-product molasses.

The other by-product is the spent beet pulp from which sugar

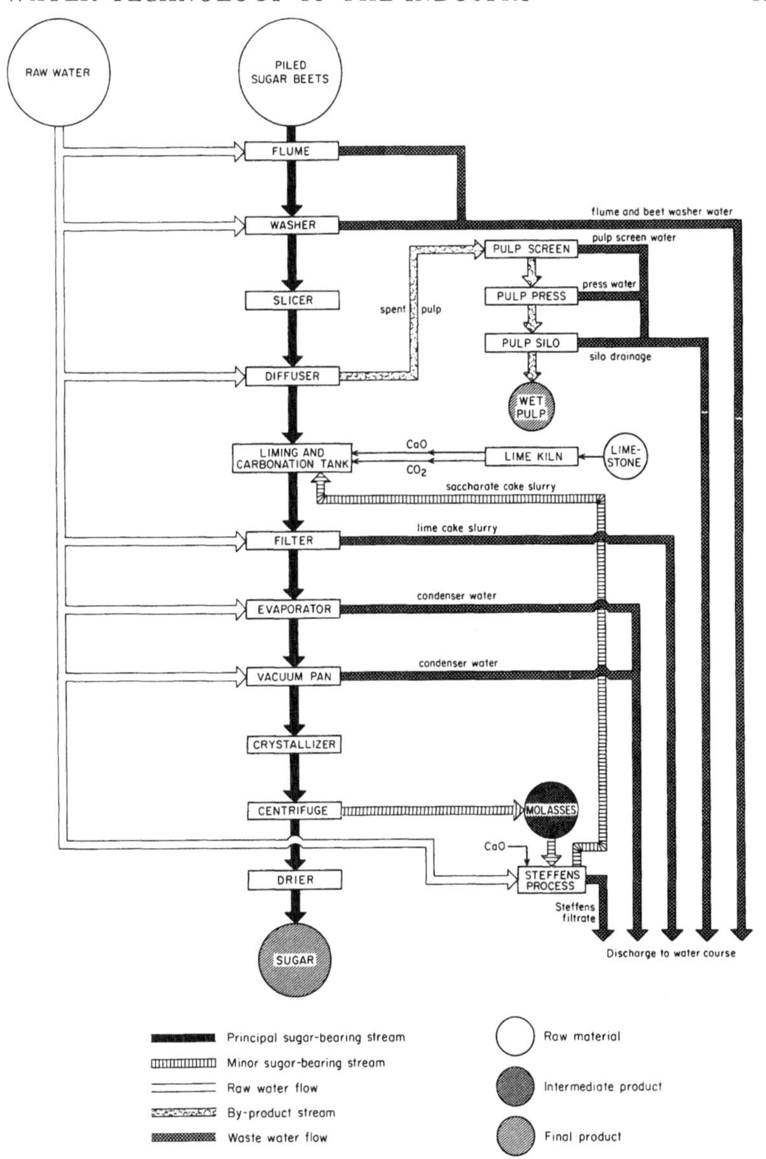

Figure 3. Materials flow in beet sugar plant with no recirculation or treatment of waste waters.

has been extracted. It is either stored in silos and sold for animal feeding as wet pulp, or more commonly is dried for sale as a dry feed component. Some or all of the molasses may be mixed with the pulp before drying.

Additional sugar extraction from otherwise by-product molasses is practiced in numerous plants by use of the Steffens process. The molasses is diluted with water, lime is added, and the resulting insoluble calcium-sugar compound is filtered from the mixture. The compound is then added as a slurry to the raw juice in the main process just prior to carbon dioxide addition, whereupon sugar is released from the insoluble calcium compound by reaction with carbon dioxide. The solution from which the calcium-sugar compound was removed (Steffens waste) is partially recycled for molasses dilution, and the balance is sewered or concentrated by evaporation prior to mixing with wet pulp.

Gross water use in each of the main processes in most beet sugar plants is shown in Table 2. Typical gross water use per ton of beets processed and per day of operation in an average plant of 2,700-ton capacity is also indicated. All these volumes

TABLE 2

Gross Water Use in Typical U.S. Beet Sugar Plant (1950)

Processing step	Gallons per ton of beets		Millions of gallons per day in typical 2,700-ton-a-day plant
	Usual range	Typical	
Beet fluming and washing	1,400–4,000	2,600	7.0
Diffusion (extraction) step (including water for conveying spent pulp)	300–700[a]	600[a]	1.6[a]
Condensing vapor from evaporators and vacuum pans and cooling of crystallizers	1,000–2,500	2,000	5.4
Lime cake slurrying and conveying	50–100	90	0.24
Dilution of molasses in Steffens process and filter washing	120–200	120	0.37
Totals (rounded)	2,900–7,500[b]	5,400[c]	14.6[c]

[a] Figures applicable to cell-type diffuser. Values for continuous diffuser are approximately one-half those shown.

[b] Range of totals shown is somewhat narrower than the 2,000- to 9,000-gallon range obtained by questionnaire (see Chapter 4), because of exclusion of unusual extremes occasionally practiced.

[c] Typical totals shown are somewhat higher than the 4,100-gallon and 11 million-gallon average values obtained by questionnaire, probably because some of the questionnaires may have excluded from the gross requirements some of the direct water reuse and also because gross water use has been materially reduced in numerous plants, especially in several large ones, since 1950, when the "typical" values shown above were determined (see also Table 3).

are the sums of water quantities previously unused and previously used for the same or other within-plant purposes.

Water and Waste Aspects of Production

A more detailed discussion of the sugar production process as directly influencing, and as influenced by, water and waste requirements and characteristics is now essayed. It will greatly aid understanding of the processes if the reader consults the flow charts (Figures 3, 4, and 5) which accompany the text.

Fluming and washing Sugar beets are taken from storage and transported into the processing plant in a stream of water flowing through a flume. From there, they enter a washer which removes any dirt which may adhere to the beets. The flume and beet wash waters may conveniently be discussed together, because, in general, they carry the same type of waste materials and are part of the same waste water system within the plant. The volume of water used for these purposes is very large and will constitute about half of the total gross use of water within the plant. Waste materials in the flume and beet wash water consist of soil particles, beet rootlets, portions of beets, stems and leaves, and a certain amount of dissolved sugar and other organic material which comes from damaged or spoiled beets. As indicated in Table 3, the biochemical oxygen demand of this water stream is substantial. It would account for about 10 per cent of the total BOD discharged by a plant in which there were no recirculation or waste recovery practices, in which wet pulp was produced and stored in silos, and which incorporated a Steffens process. The significance of the latter provisos will become clear as we follow through the various processes involved. The type of plant envisaged is shown in Figure 3. It is not representative of plants presently in the industry, but is typical of operations twenty to thirty years ago. It is useful for expositional and bench-mark purposes because it has the simplest water and waste water system. Where statements are made below about proportions of "potential" waste load which originate in various processes, the point of reference is a plant of this type.

TABLE 3

Wastes from Each Process in Beet Sugar Manufacture[a]

Waste	Waste water flow/ton beets sliced (*gal*)	BOD (*ppm*)	BOD/ton beets sliced (*lb*)	Suspended solids (*ppm*)	Suspended solids/ton beets sliced (*lb*)
Flume water	2,600	210	4.5	800–4,300	17–93
Screen water:					
bottom dump cell	240	980	2.0	530	1.0
continuous diffuser[b]	400	910	3.0	1,020	3.4
Press water	180	1,700	2.6	420	0.6
Silo drainage	210	7,000	12.3	270	0.5
Lime cake slurry	90	8,600	6.5	120,000	90.0
Condenser water	2,000	40	0.67	—	—
Steffens waste	120[c]	10,500	10.4[c]	100–700	0.1–0.7
Totals:[d]					
Using bottom dump cell	5,440[d]		39[d]		
Using continuous diffuser	5,600[d]		40[d]		

[a] Figures in the table are based on an intensive measurement program at two "typical" beet sugar plants throughout an entire campaign. One was a Steffens plant and one a non-Steffens plant. The range of values and the individual quantities shown are thus from only a small sample of the industry. They agree reasonably well with values reported on beet sugar factories in Europe. (See J. M. Henry, "The Problems of Waste Waters in Sugar Factories," in International Union of Pure and Applied Chemistry, *Re-Use of Water in Industry* [London: Butterworth & Co., 1963].) However, they are generally lower than quantities reported by certain pollution control authorities. (See Louis Parenteau, "Waste Disposal Problems at the Great Western Sugar Plants in Colorado," January 1965; South Platte River Basin Project Report, April 1966.) This Colorado program, although covering ten beet sugar plants, was of short duration during only two separate intervals. It is therefore concluded that the above table contains the most reliable data now available, but it is recognized that some plants may consistently generate more BOD and suspended solids than indicated and that all plants from time to time will generate substantially higher waste quantities because of process upsets, deteriorated beets, climate conditions during harvest, and other factors.

[b] Water-transported pulp in lieu of mechanical conveyor.

[c] Calculated on the assumption that 4.5 tons of molasses are produced and further treated by the Steffens process per 100 tons of beets. Per ton of molasses, the quantities are 2,640 gallons and 231 pounds of BOD. Steffens plants frequently process molasses shipped from non-Steffens plants, thereby altering this tonnage ratio.

[d] Totals are the water flows and BOD quantities generated in and discharged from all processes, assuming no recirculation and no waste removal from individual or combined streams.

Source: U.S. Public Health Service, "Industrial Waste Guide to the Beet Industry" (December 1950).

The most practical way to reduce water intake for the fluming and washing of beets and simultaneously to reduce the volume of waste discharged from the process is to recirculate the flume and wash waters. Recirculation of this stream is often coupled with some intervening treatment of a relatively simple type. The water may be passed through screens, a sedimentation basin or mechanical thickener, and, if recirculated in a closed cycle, is subjected to intermittent chlorination or formaldehyde treatment to control bacterial activity. However, control of bacteria can be a serious difficulty. This point is expanded later when we discuss the costs associated with various process changes. When the flume and wash water stream is fully recirculated, the plant's overall water intake demand drops by perhaps one-half. Its potential BOD load falls by about one-tenth. By these means, pollution from the flume and washing processes can be eliminated, if accompanied by suitable waste treatment, and the intake for these processes reduced to a small amount of makeup water. This option is shown in Figure 4. More typical in the industry is partial recirculation of the stream, which is also shown in Figure 4.

Water Flow Through Diffusers From the washer the beets are conducted to a scale by means of an elevator and then through a slicer where they are cut into thin ribbons called cossettes. The cossettes enter diffusers where sugar and other soluble organic materials are extracted from them with warm water. One product of the diffusion stage is a dilute juice containing about 10 per cent sugar which proceeds on to the following steps in the process. The other is beet pulp from which the sugar has been extracted.

Two types of diffusers are in use. In the older type, beet cossettes have sugar extracted from them in a battery of about a dozen large steel pressure vessels. Hot water is pumped through these, in series. Fresh water enters the diffuser containing the cossettes from which sugar has been most completely extracted, then flows to the next "oldest" batch, and so on to the diffuser most recently charged. In this semicontinuous, counter-current extraction process, the sugar concentration

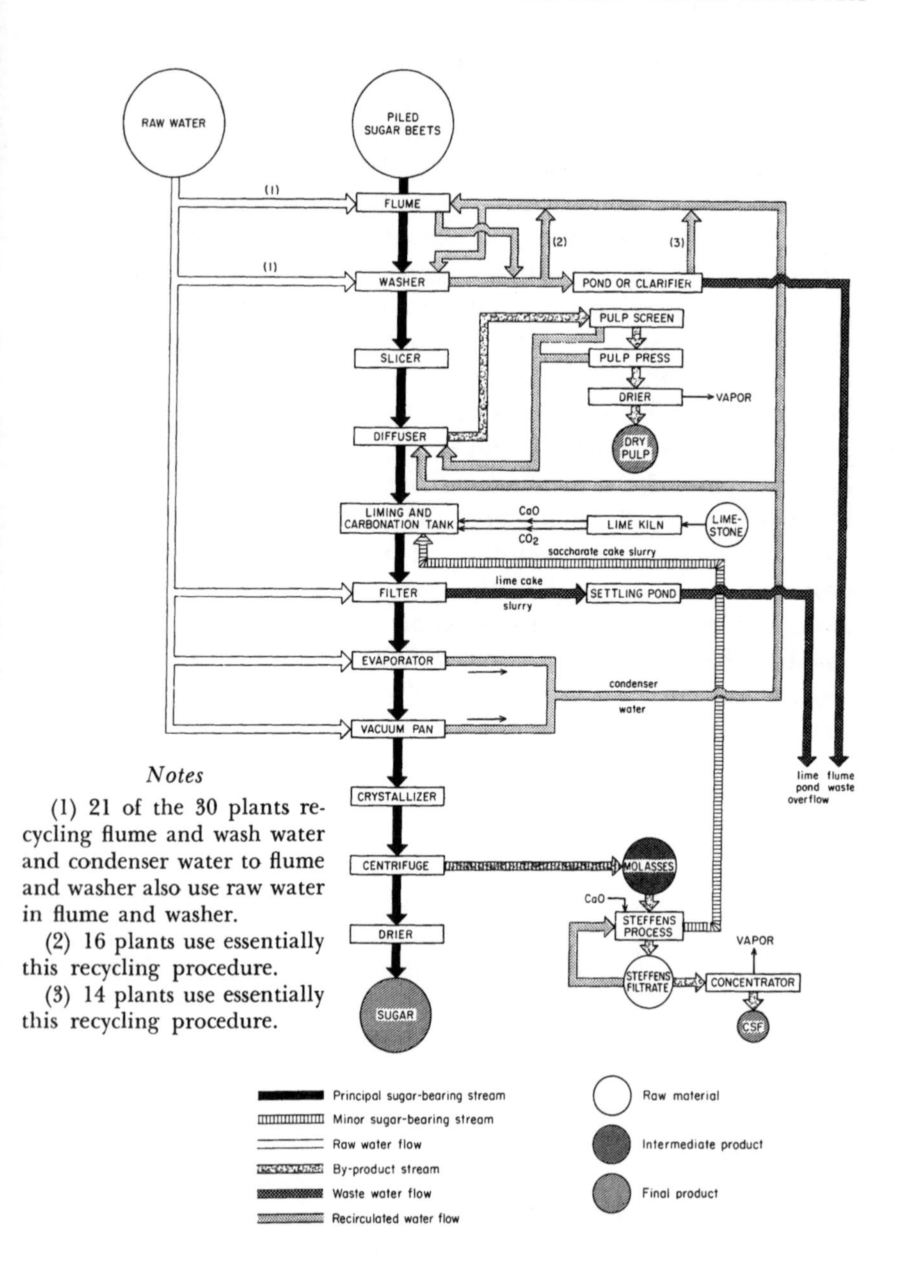

Notes

(1) 21 of the 30 plants recycling flume and wash water and condenser water to flume and washer also use raw water in flume and washer.

(2) 16 plants use essentially this recycling procedure.

(3) 14 plants use essentially this recycling procedure.

Figure 4. Materials flow in beet sugar plant with "typical" water utilization and waste water disposal pattern.

in the juice is thus built up as high as practical, whereas the residual sugar content of the beet cossettes is reduced nearly to zero. When a particular diffuser has been on stream long enough for almost complete extraction, the extracted pulp is flushed out through a bottom opening, along with enough water to facilitate flow. The diffuser is then recharged with fresh cossettes and the water flow sequence altered so that this unit becomes the last in the series.

The water used for flushing and transporting the spent pulp picks up a certain amount of very fine pulp particles, some sugar, and other organic materials. This stream passes to a screen where most of the water drains away. The pulp is then usually pressed to remove as much water from it as possible prior to drying (the usual practice) or to storage and sale as wet pulp. Water flow from the presses is usually combined with flow from the screens to form a potential waste water stream. This consists of perhaps 10 per cent of the total waste volume generated in the beet sugar plant, as indicated in Table 3. It will also contain about 10 per cent of the potential BOD coming from a plant using battery-type diffusers. Figure 3 shows the screen and press water discharged from the plant. This stream of water may be partially or nearly completely reused in the diffusion process after appropriate treatment, including further fine screening and chlorination. When this is done, waste discharge is reduced and a small amount of additional sugar is recovered.

A major improvement on the battery-type diffuser is a newer device called the continuous diffuser. In the continuous diffuser, cossettes are conducted up a large upward-sloping tube or trough by means of a slowly rotating helical screw. Upward travel is against a stream of hot water which continuously extracts sugar from the cossettes. Spent cossettes are continually discharged at the upper end of the diffuser and usually transported to the next operation by flushing with water. Separation of water from spent pulp by screening and pressing is accomplished as in plants using the older diffusion battery. The continuous diffuser lends itself more readily to the recirculation of screen and press water, because its more open system involves a better water balance, and it avoids problems associated with fine pulp suspended in the recirculated stream and gas formation in the cell-type diffusers. It also facilitates the transport of spent pulp to the presses by mechanical conveyor rather than by water flushing, thereby reducing water requirements. The continuous diffuser in current practice usually eliminates the need for external discharge of pulp screen and press water. This system, shown in Figure 4, is commonly used in existing plants.

Depending upon how it is handled, the pulp itself may cause a serious waste water problem. The pulp has considerable value as cattle feed, and one method for handling it for that purpose is to store the wet material in large silos. Drainage from pulp silos is very high in BOD and would account for roughly 30 per cent of the potential waste load from a beet sugar factory (Table 3). A commonly practiced alternative is artificial drying of the partially dewatered pulp. When the latter is done, the problem of silo drainage is, of course, eliminated as is the major waste load problem associated with it. These two methods for pulp handling are shown in Figures 3 and 4 respectively.

Further Uses of Water in Production Let us now return to the actual process of producing the sugar. The diffusion batteries or the continuous diffuser produce a raw juice which is sent on its way to two liming and carbonation tanks operated in series. In the first unit, milk of lime is added and carbon dioxide is bubbled through the limed juice. This causes the precipitation of calcium carbonate which entrains suspended solids and various nonsugar substances coagulated by the lime. The mixture is then filtered and a so-called lime cake removed from it. The remaining liquid is introduced into the second unit where more carbon dioxide is bubbled through it, this serving to precipitate the remaining lime. Filtration again takes place and more lime cake is produced. After the second filtration, the clarified juice is chemically treated to adjust alkalinity and pH.

Before following the thin juice further along its route, we wish to focus on the lime cake, which is another potential source of polluting substance. The lime cake is watered to a slurry which is comparatively easy to transport by pumping. Small amounts of sugar and larger amounts of various other organic materials remain in the slurry, which will carry over 15 per cent of the potential oxygen demand from a plant of the character previously indicated. However, it may contribute the major part of the suspended solids waste load if preliminary treatment (sedimentation) is not used. This situation is shown in Figure 3. If the slurry is not discharged directly to a stream, it may pass through a settling pond. Ordinarily this eliminates practically all of the suspended solids and produces a substantial reduction of the BOD, mainly due to settling of solids, usually to a sixth or a seventh of its former amount. After the lime slurry has dried in the ponds, it may be sold in some regions to farmers for its soil-conditioning properties, but usually it simply accumulates in the ponds until after several years it is either scraped to the sides and added to the pond walls or it is removed to a dump. Figure 4 shows ponding of the lime cake slurry which is now universal practice in the industry.

While temporary storage in ponds does reduce the BOD from the lime cake very substantially, it does not eliminate it. There is, however, a possibility of completely getting rid of this waste source by a comparatively simple (but not necessarily economical) means, should that be desired. Some plants in England and Germany do not dilute the lime cake, which is in a semidry state (50 per cent water) when it comes from the filters. Instead, the lime is disposed of in this semidry state, usually by selling it directly to farmers. Sales of the cake can be made only in areas with acid soil—this generally occurs in the more humid regions. The procedure requires mechanical conveyance of the cake at higher cost, and results in a small reduction in water use at the plant.

It is also possible to store the entire slurry flow in a lagoon and to eliminate the water by natural evaporation and seepage. The procedure, of course, results in some degree of water loss from the sugar plant. This will be in the neighborhood of 90 gallons per ton of beets processed. A number of plants practice complete containment of the lime cake slurry. Odors are sometimes a problem, particularly in populated areas, so dry disposal by conveyor and piling or burying may become more important in the United States.

Turning again to sugar production, the thin juice from the

carbonation and filtration steps contains 10 to 13 per cent sugar. The next stage in the process is to thicken or concentrate the juice. This is accomplished in multiple effect evaporators, which produce a solution containing 55 to 65 per cent sugar. A water spray is used to condense the vapors produced by the evaporators (enabling them to operate with a partial vacuum), and this gives rise to a potential waste water stream, the so-called condenser water. In comparison with the other wastes discussed, this is a comparatively minor one in respect to total BOD produced. Its oxygen demand is determined by the amount of entrainment of sugar solution in the vapor, which is largely a function of the design of the equipment employed and its mode of operation. In general, this stream carries only about one per cent of the potential BOD of a beet sugar plant, but it also has a rather high content of heat, which may or may not present a problem. The heat can be overcome by the installation of cooling towers or spray ponds, from which it is comparatively simple to recirculate the water for condensing purposes. The condenser water may be used without cooling for fluming the beets. Recirculation for fluming is the typical pattern in the industry and is shown in Figure 4. If the flume water itself is recirculated, condenser water may serve as makeup. If the condenser water could be totally recirculated, the BOD and temperature problems which might be associated with its discharge to the stream would, of course, be simultaneously eliminated. But as with all recirculated streams, the buildup of dissolved materials is a problem. If the recirculation systems are arranged so that fresh water makeup is supplied initially for condensing, and part of the condenser water is used as makeup for fluming, it is possible to keep the dissolved solids buildup to a tolerable level in both streams throughout the length of a campaign.

The evaporators have now concentrated the solution, termed "thick juice," to the point where it is about one-half sugar. Again, chemical adjustment of pH may take place, and dissolved sugar from a later process is returned to the juice at this stage. We now have what is referred to in the industry as "standard liquor." This is charged to vacuum pans (actually large evaporators) where it is boiled and concentrated and where crystals form. Evaporation is continued, with periodic addition of standard liquor, until crystals have grown to the desired size. The resulting mixture of syrup and crystals then enters centrifuges in which the sugar crystals are retained on a screen while the syrup is thrown through. Remaining adhering syrup is washed through the screen with a small spray of hot water and the wet sugar goes to a granulator for drying, cooling, and screening. Pure sugar has now been produced which is ready for packaging and distribution.

The thick syrup and the washings from the centrifuge contain much sugar in solution, a large part of which can be recovered by further concentration in a second vacuum pan. The organic and inorganic impurities in the syrup at this stage reduce the purity of the crystallized sugar, so after separation from the remaining viscous syrup in a centrifuge, this "high raw" sugar is redissolved ("melted") in hot water and added to the solution entering the first vacuum pan. Thus, additional pure sugar is obtained and impurities are largely retained in the residual molasses. This molasses is again processed by evaporation, crystallization, and centrifuging to yield a "low raw" sugar which is dissolved and added to the juice entering the high raw vacuum pan, while the molasses from this last separation finally leaves the process. Four to five tons of molasses are commonly obtained from 100 tons of beets.

The Steffens Process and Its Waste We have so far described what is known as the straight house process, common to all beet sugar factories. When most of the existing plants were built, this was, in fact, the end of the sugar-making story. The syrup which the centrifuges had cast through the screens was collected and sold as molasses. The major market was for livestock and for fermentation alcohol manufacture.

Some years ago, however, a process was invented which chemically extracts an additional quantity of valuable sugar from the molasses. This is known as the "Steffens" process and the part of the plant in which it is conducted is called the "Steffens house." The sugar content of the molasses from the straight house operation is about 50 per cent, this being about one-tenth of the sugar in the beets. Molasses is first diluted to a sugar concentration of about 7 per cent. Pulverized lime is added to this solution, forming calcium trisaccharate which is comparatively insoluble. When well operated, this process removes about 85 per cent of the sugar from the molasses. The mixture is filtered and the insoluble saccharate cake is delivered to the carbonation tank (previously described) where carbon dioxide breaks up the calcium-sugar compound by freeing the sugar and forming insoluble calcium carbonate sludge. This mixture then proceeds through the recovery process already outlined.

The solution filtered from the saccharate cake is the waste resulting from the Steffens process. It has a small volume but an extremely high biochemical oxygen demand. Indeed, this is one of the major potential polluting substances generated in a sugar factory. About one-quarter of the entire organic potential waste load may stem from this single source. In addition to its sugar content, Steffens waste contains mineral salts, principally potassium carbonate, sugars other than the common sucrose, such as manose and raffinose, and other organic compounds extracted from the beets. Figure 3 shows a sugar plant which discharges the Steffens waste directly to a water course.

Rather than discharging this highly polluting waste, most plants now reclaim it for by-products. The usual recovery method is concentration by evaporation, yielding concentrated Steffens filtrate (CSF), which is mixed with wet beet pulp prior to drying for livestock feed. In one factory, CSF shipped from several beet sugar plants is processed for monosodium glutamate (a food flavoring), for potassium sulfate, and for a residual concentrated solution for animal feed mixtures. Figure 4 shows the disposal of Steffens filtrate by concentration.

There are also other processes for recovering nearly all of the small amount of sugar remaining in the Steffens waste solution. Among them are procedures which involve osmosis and methods based on precipitation of sugar with compounds of barium or strontium. The precipitation methods are practically the same as in the Steffens process, but with the substitution of the hydroxides of barium or strontium for calcium. The principle utilized is the greater insolubility of these saccharate compounds than shown by calcium saccharate, thereby making possible the recovery of additional sugar from the Steffens waste. However, the high cost of barium and strontium compounds requires their recovery and reuse, so the precipitated carbonate resulting from the carbonation process must be converted to the hydroxides in an auxiliary process. For economic reasons the barium saccharate process has found application in only one plant in the United States and the strontium process is not used at all. It does not appear that further application of these processes will prove economical in the foreseeable future.

The processes in the "straight house" portion of all beet sugar plants are basically the same and produce identical outputs. This means the processes of the entire plant except for the facilities which may be provided for handling the by-products—pulp and molasses. Consequently, the gross water use, water depletion, and waste generation in the fluming, washing, diffusion, pulp screening and pressing, lime cake slurrying, cooling, and condensing steps are, other things equal, not affected by the product mix, be it wet pulp or dry pulp, molasses or Steffens products (additional sugar and Steffens filtrate and/or its further derivatives). In the following sections, many comparisons are made between water use in these main processing steps from plant to plant, and these are not biased by the final product mix.

Pulp drying, however, as contrasted with wet pulp production, although not affecting gross water use, does greatly reduce waste generation and discharge by virtue of silo drainage elimination. The product mix (wet versus dry pulp) thus influences total waste quantity.

The Steffens process increases sugar production per ton of beets handled by about 10 per cent. It increases gross water use only slightly (about 2 per cent in Table 2) and water withdrawal practically not at all, so that comparisons between plants as a whole in terms of gross water use or depletion are not substantially biased by the presence or absence of a Steffens facility. Such comparisons are also made in the following chapter. The amount of sugar produced per unit of water used is, of course, increased slightly if a Steffens plant is operated.

Implications of Process Changes for Saving Water and Reducing Waste Loads

We have gone through a description of the technology of this industry with a special view to its use of water and generation of waste streams. We have pointed to some of the major ways of reducing or even eliminating waste streams.

One conclusion which can readily be drawn at this point is that it is perfectly feasible to design plants which have processes and recirculation patterns that produce no external waste load at all. Such a plant is shown in Figure 5. This is by no means to imply that it would always be desirable to design plants this way. While most of the possible process changes involve additional sugar recovery or the winning of other valuable products, many of them imply higher costs than would have to be incurred if water supply were available in unrestricted quantity and wastes could be discharged from the plants in an uninhibited fashion. If these conditions prevailed, flume water would be discharged to the stream as probably would pulp screen and press waters. Whether or not driers would be substituted for silos as a means of processing pulp would depend upon market conditions. One may hypothesize too that it would almost always involve lower internal costs to the factory to discharge the lime cake slurry directly into a stream. The same is true of condenser water. Whether or not the recovery of monosodium glutamate and potash from the Steffens house waste would be undertaken is largely a function of the market for these products and the availability of appropriate processing plants. Economies of scale in the production and marketing of the products recovered from the Steffens house waste make it highly uneconomical for individual beet sugar plants to attempt to install these processes. Accordingly, while plants situated close to by-product recovery facilities may readily increase their internal rates of return by shipping the Steffens wastes to them, others not so favorably situated might need to incur a net cost in order to handle this waste stream.

For all these reasons, waste reduction and water intake diminution come only at a cost. It is one of the major purposes

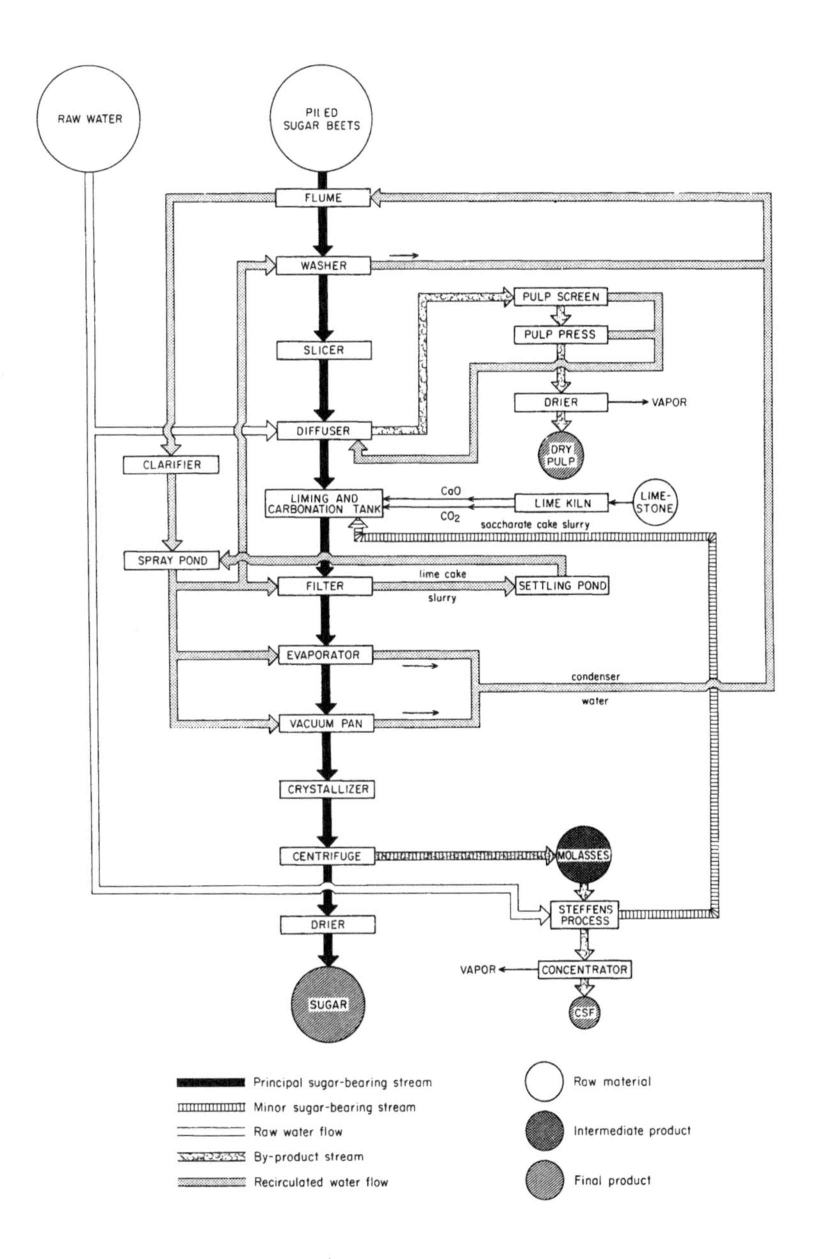

Figure 5. Materials flow in beet sugar plant with full recirculation and no external discharge of waste loads.

of this study to establish functional relationships between these water use variables and the net costs of altering them. With the understanding of technology and costs, which the sugar plants have helped us to gain through their impressive response to our questionnaires, and through special studies which they have performed for us, we can provide guidelines concerning probable processes, and the attendant costs, which would be instituted under various water environmental conditions—primarily water costs and restrictions on waste discharge.

Other Methods for Reducing Waste Loads

The exclusive attention which has thus far been given to process changes should not lead the reader to suppose that there are not alternatives for reducing waste loads. The relevant ones, once a waste stream has left a particular process are as follows:

1. *Mechanical Separation of Suspended Matter* A large fraction of the waste materials produced in beet sugar plants is in the form of solids carried along in the several waste streams. These can be rather easily and often cheaply removed by (a) screening, and (b) settling or sedimentation.

a) Screening. Trash, leaves, small pieces of beets and the like can be separated from flume and beet washer waste water by passing the entire flow through a grating and a fine-mesh stainless steel or bronze screen of usually a rotating, vibrating, or traveling type. Mechanical vibrations or water sprays remove the solids from the screen for land disposal or admixture with by-product beet pulp. Fine pulp is removed in similar fashion from pulp screen and press waters.

b) Settling and sedimentation. Fine soil particles and other matter suspended in flume water, and precipitated lime suspended in the lime cake slurry, may be settled in ponds or in mechanical thickening tanks. In addition to the resulting suspended solids removal, this process accomplishes substantial BOD reduction in the water overflow to ultimate discharge. All beet sugar plants employ this process for treating at least a portion of the liquid wastes produced.

2. *Treatment of Waste Water in Plant Using Biological Processes of Waste Degradation* The relevant biological treatment methods are similar to those used for treating municipal sewage. They subject the waste stream to biological degradation in a controlled environment within a treatment plant.

There are two major types of biological treatment plants for sewage containing decomposable wastes, (a) the one referred to as a trickling filter, and (b) the activated sludge plant. Although neither process is regularly used for treating beet sugar factory wastes, a number of experiments have been conducted with one of them. In a German experiment reported by Sierp[1] the effluent from the plant was treated directly on a trickling filter. This required large-scale dilution of the effluent. The diffuser and press waters had to be reduced to a sugar content of less than one-tenth of one per cent. Even at this low concentration, the trickling filter could not be heavily loaded, and it took some period of time for the biological cultures to develop to an extent where treatment became effective. Under these circumstances, direct treatment of sugar factory effluents on trickling filters appears to be extremely costly. In another German experiment also reported by Sierp,[2] effluent from a sugar factory was stored in a lagoon which had sufficient capacity to permit treatment on a trickling filter to proceed beyond the campaign, indeed, throughout the year. In this instance comparatively good results were obtained with a much more modestly sized plant, and eventually it became possible to load the trickling filter with the undiluted waste water from the lagoon. Of course, the combined investment in lagoon and trickling filter could be quite large. As with all other external treatment devices and procedures for reducing the pollutional effect of effluents, the investment in lagoon and treatment facilities can be substantially reduced if internal recirculation and waste recovery processes are instituted. Under certain conditions where a high quality effluent from the plant is justified, a combination of various measures internal to the plant plus lagooning

1 Friedrich Sierp, *Gewerbliche und Industrielle Abwässer* (Berlin: Springer Verlag, 1959), p. 285.

2 *Ibid.*, p. 286.

and treatment on trickling filters may be the optimum arrangement. It is, of course, not possible to compare alternatives until after we have developed cost figures for them.

3. *Temporary Storage of Wastes with Either Continuous or Programmed Discharge* If discharge from a pond is continuous, wastes will be degraded biologically to varying extents depending on the period of retention in the waste pond. Intermittent or programmed discharge can take advantage of high streamflows or other favorable conditions for waste discharge. Since programmed discharge requires some period of waste storage, biological degradation forces will of course be operating. Ponding is very important in the industry and is discussed in considerable detail in following chapters.

4. *Storage of All Wastes, Plus Evaporation and Seepage* Where land is available, large lagoons may be employed for containment of all beet sugar factory wastes. Additions of waste waters during the campaign are balanced by evaporation and underground seepage continuing throughout the year. Decomposition of organic material also occurs, sometimes creating odor nuisance. The practical limitations to this waste disposal method are usually the cost of large acreages required for complete waste containment.

5. *Spreading of Waste Waters over Land* The goal of landspreading may be infiltration of waste waters and the degradation of organic wastes in the unsaturated soil zone, or degradation of organic materials by permitting the waters to flow through a grassy field.

Several experiments have been performed upon the land disposal of beet sugar waste waters. The Emschergenossenschaften Essen, Germany, for example, sponsored an experiment involving a permanent system of underground pipes. The costs of facilities of this character, of course, depend heavily upon land characteristics, distance to the land, costs of electric power, and so forth. Unfortunately, the seasonal period of sugar production is out of phase with the growing season of crops, and consequently direct irrigation of crops is not in question. Other German experiments have combined ponds, in which fermentation takes place, with a very high rate of sprinkler irrigation of land. The water draining from these fields is said to contain relatively little residual BOD. A possibly difficult problem arises, since in order to avoid conditions of aesthetic nuisance from the fermentation ponds, it is necessary to control the process chemically and with great care.[3]

An interesting experiment involving broad field disposal of sugar beet wastes was conducted in this country by Ralph Porges and Glen J. Hopkins.[4] In this study, a liquid waste flow amounting to more than 2,400 gallons per ton of beets processed was spread upon a natural grassland treatment field. The flow amounted to 4.3 cubic feet per second with an average of 483 parts per million BOD. Of this, 3.5 cubic feet per second with an average BOD value of 158 ppm were returned to the drain ditch as effluent from the field after an average of 14½ hours of detention in the treatment plot. Treatment on the field resulted in a BOD reduction of 73 per cent and suspended solids removal of over 99 per cent.

Industry Practice in Handling Waste Loads

Neither land disposal nor treatment in conventional treatment plants has become common practice in the beet sugar industry. The former requires large acreages near the plant. These are often unavailable and the latter involves technological and economic problems not yet fully solved. An exception may be in Germany where discharge of factory effluents to public sewerage systems and subsequent treatment in municipal treatment plants appears to be rather common. Two plants in the United States use this method also. When mixed with domestic and possibly other industrial waste waters, such treatment appears quite feasible and probably is rather economical. In the United States, however, this would only infrequently be a feas-

[3] *Ibid.*, pp. 283–87.

[4] Porges and Hopkins, "Broad Field Disposal of Beet Sugar Wastes," *Sewage and Industrial Wastes*, Vol. 27, No. 10 (October 1955), pp. 1160 ff.

ible way of handling the waste waters, because plants are usually located in rural areas.

Much more common is the lagooning of various waste flows either explicitly for waste treatment or for programmed discharge. Thus lagooning of the lime cake slurry and of the Steffens house waste—where not otherwise disposed of—is common practice. Description of the extent of and costs of lagooning these wastes and analysis of the future potential of the practice are presented in subsequent sections. In some instances other wastes are lagooned, such as pulp screen and press water and flume water.

Storing the entire potential waste water flow from a beet sugar factory requires large amounts of storage space. For example, a plant processing 2,000 tons of beets a day during the campaign with a gross water use of 5,000 gallons per ton of beets processed would, in the absence of recirculation, require 1,500 acre-feet of storage to provide a retention time of six weeks, and about 4,000 acre-feet to contain the waste water from a full 100-day campaign.

Waste waters other than lime cake slurry and Steffens waste are occasionally lagooned in separate ponds as an intermediate stage in a recirculation process. Water from the lagoon is then drawn upon for flume water, condenser water, and for certain washing and flushing uses not requiring high quality.

Other possible combinations of measures affecting water intake or waste discharge from sugar plants will be outlined in the subsequent sections when actual empirical information on these practices and the associated costs is presented. The objective here has been merely to give an overview of the range of alternatives which are possible under current technology.

4

Water and Waste in U.S. Plants

Detailed information on water utilization in the beet sugar industry was obtained by circulating two questionnaires to the twelve beet sugar companies in the United States. These companies operated fifty-eight manufacturing plants in 1962,[1] the year for which current data were requested, and they accounted for all the beet sugar manufactured in the United States that year.

The Questionnaire

Response to the first questionnaire was unanimous, but replies came in varying degrees of completeness. Examination of the results showed that additional information was needed, so a supplementary questionnaire was sent to each of the companies and plants involved in the first survey. Fifty-three of the fifty-eight plants replied to this questionnaire, and the data supplied were reasonably complete.

Copies of the questionnaire are shown in Appendixes B and C. The first is patterned after one which was distributed to all U.S. manufacturers in 1960 by the National Technical Task Committee for Industrial Wastes.[2] Modifications and additions were made to permit procurement of more precise information on this particular industry. The principal data requested included beet tonnage, water intake, details on water recirculation, water and waste treatment processes and their associated costs, discharge quantities and qualities, composition of raw water and waste waters before and after treatment, source of water supply, types of reuse and recirculation, plans for con-

[1] In 1964, sixty-two plants were in operation.

[2] Results of this survey are reported in *Water in Industry,* National Association of Manufacturers and Chamber of Commerce of the United States, January 1965.

struction of new water and waste facilities, the disposal of wastes by means other than discharge to streams, and reductions in waste production through recirculation. In the supplementary questionnaire, primary emphasis was placed on comparison of beet tonnage and water use in 1949 and 1962. Specific information was requested on new water recirculation processes and on the handling of wastes in ponds of various types.

The questionnaire responses provided data for all plants on water intake, beet tonnage, sugar production, recirculation quantities, and waste treatment processes used. In most instances, data were also available on the capital costs and operating costs of intake water treatment processes, if any were required, and on the waste treatment processes, if any were used. Data on water recirculation in many plants were not completely self-consistent, possibly due to differences in interpretation of the questionnaire by the respondents.

Data on composition of waste waters were considerably less complete than other information supplied. Some companies apparently have very little information on this subject. Others have analyses of process and plant discharge streams, in respect to BOD, suspended solids, and total solids.

The excellent cooperation of the industry merits special comment. The unusually high response to the questionnaire and much helpful cooperation from company personnel is interpreted by the authors to indicate exceptional interest and concern by this industry in its water problems.

Water Withdrawal or Intake

The beet sugar plants surveyed range in daily capacity from 1,100 tons of beets to 6,700 tons. The average capacity is about 2,700 tons per day. Since many plants cluster around this size range, we define a 2,700-ton plant as typical and use this term in the following exposition.

Annual tonnage covers a somewhat wider range because of varying climatic factors in different parts of the country and the resulting lengths of production runs, or "campaigns." Thus, the minimum annual plant tonnage processed is about 94,000 tons and the maximum approximately 1,060,000 tons. The average annual throughput is about 325,000 tons of beets, from which 40,000 to 50,000 tons of sugar are typically produced. Thus, about 0.12 to 0.15 ton of sugar is obtained from a ton of beets.

Variations in annual beet tonnage processed by a factory operating at a constant daily throughput are due primarily to differences in the year-to-year harvest of beets. These differences are, in turn, caused mainly by weather variability and to a smaller extent by changes in total acreage planted. An annual variation of 20 per cent in beet tonnage is commonly experienced. The much larger annual tonnage of California plants having the same daily capacity as plants in other states is the result of two beet harvests per year in that state.

Fluctuations in sugar yield per ton of beets are also attributable to the weather. In general, greater sunshine and warm autumn weather favor higher beet sugar contents, whereas wet and cloudy conditions or early freezing weather result in lower sugar yields. With beets of good quality, the portion of their sugar content extracted and recovered for sale is relatively constant, plant-to-plant and year-to-year. However, beet deterioration due to prolonged storage or excessive physical damage and occasional process breakdowns are accompanied by reduced yields. Between one year and another, these factors have a combined effect of as much as 15 per cent in the amount of sugar produced from a ton of beets in a particular plant. Between plants of the same type (that is, non-Steffens), there may be another 15 to 20 per cent variation in the product/beet ratio, mainly because of regional differences in beet sugar content. Although sugar recovery in a Steffens plant is usually about 10 per cent higher than in a non-Steffens plant, there appears to be very little systematic difference in production efficiency (sugar product per unit of sugar in beets) due to other manufacturing variables such as plant capacity, age, and operating procedure.

Water withdrawal per plant covers a very wide range, from a low of 430,000 gallons per day in one plant to a high of 13 million per day. Average withdrawal for the industry, per plant, was 6.4 million gallons per day. On the basis of water withdrawal per ton of beets, the quantities ranged from a low of 270 gallons to a high of 5,250 gallons, with the typical plant (2,700 tons of beets per day) withdrawing 2,600 gallons per ton of beets processed. The total water intake of all plants divided by total beet tonnage processed is 2,360 gallons per ton. The 2,600 figure agrees with the U.S. Census figure for 1959 shown on page 14. Average water withdrawal is thus about 18,000 gallons per ton of sugar produced, or 9 gallons per pound.

The exceptionally wide range in water intake through the industry illustrates the degree to which water conservation is possible. Typical gross water use, including recirculation, is roughly 4,100 gallons per ton of beets; this figure would be the approximate average water intake if no recirculation were practiced. The U.S. Census figure for 1959 is 4,400 gallons gross use per ton. It is seen that by recirculation, withdrawals have been reduced to an average less than two-thirds of this figure, in several plants to less than one-fourth, and in at least one plant to about one-fifteenth of average total use. This indicates that availability of large quantities of fresh water for intake purposes is not a determining factor in beet sugar plant location and operation. Small supplies of water can be compensated for by higher investment in water utilization equipment.

To ascertain the principal reasons for the large variations in water intake, gross use, and recirculation, the effects of various beet sugar plant features on trends in water usage have been appraised by two procedures, the results of which are presented in Tables 4a and 4b. Table 4a is based on the results of a multiple regression analysis of questionnaire data, as explained in Appendix A. Table 4b contains somewhat more qualitative conclusions drawn from the general characteristics of the processes used in beet sugar manufacture.

Significant differences in water withdrawal for plants in various regions were observed in the questionnaires and are summarized in Tables 4a and 4b. Nearly all plants in California (Table 4a) use comparatively small amounts of intake water. The average in California, for example, is about 1,300 gallons per ton and in Idaho, 2,450. In the Midwest, including Ohio,

TABLE 4a

Effects of Various Beet Sugar Plant Features on Trends in Water Usage, as Determined by Regression Analysis of Questionnaire Data

Plant-to-plant differences	Gross water use per ton beets, x_3	Gross flume water use per ton beets, x_{11}	Intake per ton beets, x_{12}	Recirculation per ton beets, x_4	Recirculation as per cent gross use, x_5
Larger plants, x_1	[a]	lower	lower[b]	lower[c]	lower[c]
Newer plants, x_7	lower	[a]	lower	higher	higher
Steffens v. non-Steffens plants, x_8	[a]	higher	[a]	higher[b]	higher[b]
Continuous v. cell diffusers, x_9	[a]	[a]	[a]	higher[b]	higher[d]
Lagooning v. no lagooning, x_{10}	lower	[a]	lower[e]	higher	higher
California v. other states, x_2	lower	[a]	lower	higher	higher

Interrelationships between various water uses:

Plants with higher gross water use tend to employ higher quantities of recirculated water.
Plants with higher gross water use also tend to employ higher per cent recirculation.
Plants using more water for purposes other than beet fluming and washing tend to employ higher per cent recirculation.

Note: x_1 to x_{12} are symbols for the variables, as defined in Appendix A.

[a] Indicates no statistically significant relationship.

[b] Relationship found, but degree of confidence only 80 per cent to 90 per cent.

[c] No statistically significant relationship for all plants, but if California plants are excluded there is a statistically significant negative relation between recirculation and plant size.

[d] Possible relationship, but confidence only 50 per cent to 60 per cent.

[e] More significant for small plants than large plants.

TABLE 4b

Effects of Various Beet Sugar Plant Features on Trends in Water Usage, as Determined from General Characteristics of Processes

	Gross water use per ton beets	Intake per ton beets	Recirculation per ton beets
Plant-to-plant differences:			
Less abundant water supply	lower	lower	same or higher
More stringent waste restrictions	slightly lower	lower	higher
Dry pulp by-product v. wet pulp	no difference	no difference	no difference
Year-to-year differences:			
Wet v. dry harvest	higher	higher or same	same or higher
Lower cooling water temperature	slightly lower	slightly lower or same	same or slightly lower
Lower beet sugar content	no difference[a]	no difference[a]	no difference[a]
Higher tonnage processed, i.e., longer campaign	no difference	no difference	no difference

[a] If these water quantities were per ton of sugar produced rather than per ton of beets handled, they would be higher when beets have lower sugar content.

Michigan, Minnesota, and Iowa, the average is about 2,200 gallons per ton. The largest withdrawals are in the mountain states where the average intake is approximately 3,150 gallons per ton of beets. These differences reflect the variations in water recirculation and conservation practices throughout the industry, which are in turn mainly due to limitations on waste discharge imposed or suggested in the various states (Table 4b) and in some cases to scarcity and/or cost of intake water supply.

Another significant fact, to which we shall return later, is the change in water withdrawals during the past fifteen years. In the supplementary questionnaire, data were requested which permit determination of unit water withdrawals in 1949 and in 1962. In the majority of plants, daily beet tonnage increased while total daily water withdrawal increased by a smaller factor or, in some instances, decreased. For the industry as a whole (based on 58 plants in 1962 and on 43 of the same 58 in 1949), the 1949 average withdrawal was about 3,180 gallons per ton of beets whereas the 1962-63 figure is 2,360 gallons.[3] This represents a decrease in water intake per ton of beets of over 25 per cent. The principal reasons for this decrease are discussed below.

Results of the multiple regression analysis of the data, shown in Table 4a, also reveal significant relationships between the size of plant and the age of plant on the one hand and water withdrawal per ton of beets on the other; the newer and larger the plant the less water intake per ton of beets. This is apparently due to the increasing importance of water and waste planning in modern beet sugar plant design and to some economies of scale in the construction and operation of water handling facilities.

Gross Water Use

Gross water use is the sum of all quantities supplied to all processes in the plant, and equals intake plus recirculation. Although intake can vary over an extremely wide range from plant to plant—mainly because of differences in recirculation—gross use must be more uniform if processes are reasonably similar. In beet sugar manufacture, there is greater uniformity in gross use than intake, but a fourfold variation is observed in the questionnaire results (see Table 2). This variation is due mainly to differences in water using practices and, to a lesser extent, to differences in the design and operation of the plant. Gross use ranges from a low of 2,080 gallons per ton of beets to a high of 8,650 gallons, with an average of about 4,100 gallons per ton. By far the largest factor responsible for this variation is the use of water for transporting and washing beets. As can be deduced from Tables 4a and 4b, flume design, the amount of soil or dried mud clinging to the beets, washer design and operation—all affect this use of water. There may also be substantial differences in the quantity of condenser water em-

[3] There were 82 plants operating in 1949, but only 56 were still in operation in 1962. Twenty-six plants were closed or dismantled in the intervening period. Since the discontinued plants tended to be smaller, older, and less equipped for water recirculation than those kept in use, the average withdrawal by *all* plants in 1949 would be somewhat higher than the 3,250-gallon average.

ployed due to temperature differences in raw water supply and in plant-to-plant operating conditions. Table 4a shows that there are no statistically significant differences in gross water use in plants employing continuous or batch diffusers and in Steffens or non-Steffens plants. Factories using lagoons for waste disposal tend to have lower gross water requirements than other plants and, in general, the California plants show lower gross water use.

Although *withdrawal* has been observed to be lower in newer and larger plants, *gross use* is less affected by these variables. The lack of dependence of gross use on plant size is due to the similarity of the processes employing large quantities of water in plants of all sizes. Table 4a does show, however, that newer plants tend to have somewhat lower gross water use and that larger plants usually employ less fluming and washing water per ton of beets handled. But the gross use of *all* water does not appear statistically related to plant size.

Seasonal differences in gross water use (Table 4b) are due mainly to variation in raw material quality, that is, the amount of foreign material such as soil and trash which must be washed from the beets, and in the temperature of the condenser water, colder water permitting slightly lower use. Per ton of sugar produced, however, Steffens plants show slightly less gross use and intake than non-Steffens because of the higher sugar yield for only a little more water.

Recirculation

The figures on recirculation were supplied in two ways. Indicated first, was the amount of additional water withdrawal that would have been required if no recirculation had been employed in the plant. Second, the quantities of water which were recirculated, with or without treatment, were separately stated. Some inconsistency was noted between these figures in a number of instances, but most of the respondents showed that the sum of treated and untreated water which was recirculated equaled the additional water which would have been required if no recirculation had been employed.

As expected, the quantities of recirculated water covered a wide range. The average for the entire industry, including those plants which did no recirculation, was 4.7 million gallons per day. Comparing this figure with withdrawal of 6.4 million gallons per day, we see that the "average" plant uses a total of about 11.1 million gallons of water per day, of which about 60 per cent is "new" water withdrawn from streams or wells, and 40 per cent is recirculated. Per ton of beets, the figures are 2,360 gallons of intake water and 1,750 gallons of recirculated water. Recirculation practice ranges from nothing in those plants using intake water for all purposes, to 85 per cent of gross use in one plant and 90 per cent in another. There appears to be very little direct relationship between recirculation and gross use. On the contrary, gross use is the more uniform quantity, whereas recirculation and intake are reciprocally related. Table 4a shows higher recirculation in newer plants, in Steffens plants, in factories with continuous diffusers and waste lagoons, and in California plants. Apparently the lower recirculation in larger plants is due to the reduced total gross use in these installations.

Data on water recirculation show considerable variation in treatment practice. In most cases, some water is treated and the balance untreated, prior to recirculation. In general, condenser water is reused without treatment, usually for fluming and washing beets. When flume water is recirculated, it is usually given simple treatment by screening and, in many cases, by settling or sedimentation. The more complete recirculation is, however, the more sophisticated treatment must be. Pulp-press water may be screened prior to recirculation to continuous diffusers. Other differences in the treatment of water streams prior to their recirculation are apparent.

Water Sources and Treatment Processes

Information was requested in the questionnaires on the sources of water for each plant, whether from wells or from rivers or lakes, and whether the supply was municipal, company-owned, or from a private water company. By far the largest supply for the industry is from rivers, but nearly one-third of the plants supply all or part of their water requirements from wells, usually company-owned. Many of the plants using only river water treat the supply by screening and settling and for at least portions of the supply, by softening (for boilers) and chlorination (for water used in actual extraction of sugar). Numerous plants using river water as a main supply also employ a small amount of treated municipal water for operations involving direct contact with sugar. Since this source is already potable, no further treatment is ordinarily required in the sugar plant.

Limited data supplied by some of the respondents show that raw water used in the plants is of reasonably good quality even before treatment. The BOD in several river water analyses ranged from two to eight parts per million, negligible in comparison with the sugar plant effluent. Most of the solids are of the dissolved variety, suspended solids usually being in the range of twenty to sixty parts per million.

Most of the well water supplies need very little treatment, but occasionally settling and softening are employed. Chlorination is also used in a few plants employing well water.

Companies show a wide range of intake water treatment costs. Part of the variation is due to differences in processes. Some plants use no treatment whereas others carry out extensive treatment of contaminated and turbid river water. The capital costs of water treatment facilities range from zero to about $50,000, with a typical cost for settling, screening, and softening equipment in the $5,000 to $15,000 range.[4] Annual costs of operating these facilities, exclusive of capital charges, also vary widely, from a few hundred dollars to a few thousand, with an approximate average of $2,000. It appears that water treatment is usually not an important cost item in beet sugar manufacture.

Further emphasis may be given to this point by comparing entire plant costs with water treatment costs. Plant investment (present costs of a large new plant) is about $4 million per 1,000 daily tons of beet processing capacity. A typical 2,700-

[4] These figures represent replacement costs as of 1962.

ton plant would thus require an investment of slightly over $10 million. A raw water treatment facility in the $15,000 cost range thus represents only 0.15 per cent of plant investment. As to operating costs, a typical annual total, exclusive of capital charges, for a 2,700-ton plant (excluding the cost of beets) may be taken as $1 million. An average annual water treatment cost of $2,000 thus comprises only about 0.2 per cent of operating costs. Beets to supply the plant would cost about $4 million. Thus, water treatment would amount to only about 0.04 per cent of all annual costs exclusive of capital charges.

A clear-cut relationship between water treatment costs and quality of the intake water does not emerge from the data. One thing that is clear, however, is that the constituents of intake water that now cause problems and costs are not associated with upstream waste discharge. The processes used to treat intake water are practically always sedimentation and softening. The former is made necessary by suspended solids of natural origin. The latter is a consequence of natural hardness.

What this means from the point of view of public policy is that, unless quality characteristics not yet experienced come to influence the costs of water preparation in this industry, pollution control practices will yield little or no benefit to it. The quality characteristics of the water important to the beet sugar industry are at least as yet not influenced in a significant way by waste discharge. Should waste disposal importantly begin to affect salinity, hardness, or suspended solids in the water courses of the sugar production areas, a small amount of benefit might accrue to the beet sugar industry from control practices.

Depletion or Consumptive Use of Water

The quantity of plant effluent before treatment from a beet sugar factory approximates the amount of intake water, net evaporation to the atmosphere being roughly equal to the water content of the beets. This relationship appears to vary in moderate degree, however. Henry[5] states the volume of waste water

[5] J. M. Henry, "The Problems of Waste Waters in Sugar Factories," in International Union of Pure and Applied Chemistry, *Re-Use of Water in Industry* (London: Butterworth & Co., 1963).

to be *greater* than withdrawal (an increase of 96 gallons per ton of beets). The U.S. Census[6] shows the waste volume to be about 200 gallons *less* than withdrawal. However, this estimate includes losses occurring in treatment processes such as ponding.[7] Calculations based on a 75 per cent water content of the beets, the small water content of molasses and other by-products, and a rough average plant evaporation based on reported fuel consumption (the latter showing about 1 ton of water evaporated per ton of beets) indicate an average water volume decrease, prior to separate waste treatment, of 50 to 100 gallons per ton of beets processed. This is 2 to 4 per cent of intake, or 1.2 to 2.4 per cent of gross use. Estimating a 3 per cent loss based on withdrawal, the total volume of plant effluent before treatment is about 360 million gallons per day (compared with 370 million withdrawal) from 58 plants, an average of 2,300 gallons per ton of beets. As shown below, return surface flow to streams is considerably less than this because of extensive ponding and lagooning of wastes, with resulting evaporation and seepage.

The estimate of water losses given in the previous paragraph is based on the assumption that waste water is discharged about at the temperature of the receiving stream, that is, that condenser water is cooled in the plant either by ponds or cooling towers. If the water is discharged at a higher temperature, there will tend to be less loss in the plant but more loss in the stream. The reason is that in both cases cooling occurs largely by evaporation. The best available evidence indicates that the overall loss will be about the same whether the cooling occurs in the plant or in the stream.[8]

Since evaporation of water in the various processes is generally a little greater than the water content of the incoming beets, there is a slight decrease in the flow of water prior to any waste treatment. All plants employ settling ponds and some

[6] Recall the difference between the Census definition and ours, discussed in Chapter 2.

[7] U.S. Bureau of the Census, *Census of Manufactures: 1958*, "Industrial Water Use, 1959," Bulletin MC58 (1)-11 (Washington: U.S. Government Printing Office, 1961).

[8] Hans H. Landsberg, Leonard L. Fischman, and Joseph L. Fisher, *Resources in America's Future* (Baltimore: The Johns Hopkins Press for Resources for the Future, Inc., 1963), p. 816.

use large holding lagoons, so there are additional evaporative losses to the atmosphere. The foregoing U.S. Census estimate of an average 200-gallon flow decrease per ton of beets represents the sum of within-plant evaporation, pond evaporation, and pond seepage. On this basis, in a typical 2,700-ton plant, 0.54 million gallons of intake per day would not appear as discharge; for the entire industry this would be about 32 million gallons per day.

There are indications that the actual difference between intake and discharge in beet sugar manufacture is substantially higher than estimated above, mainly because of the large increase in ponding of wastes since the 1958 Census. Table 7, p. 63, shows considerable growth in use of ponding since 1960, and some of the installations between 1950 and 1960 were made after the Census estimates. The fact that only five plants with no surface discharge withdraw a total of 25 million gallons per day shows that the whole industry must be consuming much more than the above 32 million total.

Statistics on water losses from settling ponds and holding lagoons in beet sugar plants are not adequate for a good appraisal of these total quantities. They are even less useful in distinguishing between underground seepage, which generally reappears downstream, and evaporation, which is not reusable elsewhere. From partial data on pond area, however, some estimates of evaporation losses may be made; and information on waste water flow rates entering and leaving the ponds and lagoons of some plants permit approximations of combined evaporation and seepage.

The total pond acreage reported by 25 plants having a combined daily capacity of 71,800 tons of beets is 2,350 acres. At an estimated average evaporation rate during the autumn season of 3 inches per month, total losses by evaporation from ponds in these plants should be about 6 million gallons per day, corresponding roughly to an average of 100 gallons per ton of beets.[9] Assuming the same extent of ponding in the entire industry, water disappearance by evaporation from ponds would

[9] This does not take account of water remaining in periods after the campaign and associated evaporation. The estimate is therefore somewhat on the low side.

be about 15 million gallons per day. Addition of approximately 75 gallons net evaporated by heat additions within the plant brings the total evaporation to 175 gallons per ton of beets or about 28 million gallons per day.

The combined evaporation and seepage from the ponds and lagoons of 35 plants having a daily throughput of 102,000 tons of beets is found to be about 60 million gallons of water per day. This is equivalent to about 600 gallons per ton of beets and at the same ratio is estimated to be 100 million gallons per day in all 58 plants. The figure, of course, varies widely from plant to plant, being near zero in some and as much as the total withdrawal in others.

It is seen from these rough estimates that evaporation losses are a comparatively small fraction of withdrawals in the beet sugar industry (6–7 per cent) and that underground seepage averages 15–17 per cent of withdrawal. Since evaporation is not a significant fraction of streamflow and since seepage subsequently reappears downstream, beet sugar manufacture has a very small effect on the quantity of water in the nation's streams. It is evident that water disappearance is seldom a limiting factor in plant location or operation.

Waste Waters

General Aspects Liquid wastes, as previously mentioned, come from several steps in the manufacturing process, and as a result these process effluents are of widely varying composition (see Table 3). The questionnaires show that there is an extremely wide range in the nature of the wastes and in the techniques used in handling them. Treatment before final discharge ranges from none at all to some form of treatment of the entire plant effluent. The surface discharge of waste water to streams ranges from the total intake volume to essentially zero in some plants employing extensive recirculation and/or evaporation and seepage ponds for all the wastes.

Waste Quantities The questionnaire asked for the quantities of liquid discharged to surface water courses and to other receivers, subdivided into treated and untreated wastes. Interpretation of the questions by the respondents and of the answers by the authors was difficult, with the result that some ambiguity and inconsistency were evident on some matters. A supplementary questionnaire was therefore used to clarify the doubtful points.

The most common waste handling practice is the discharge of a portion of the waste water directly to rivers without treatment and the delivery to a settling pond of most of the remaining portion for temporary holding, settling, and then discharge to the stream. The effluents containing the highest waste concentrations, such as the lime cake slurry and the Steffens filtrate (where it is not sent along to another plant for further processing), usually go to a settling pond, whereas condenser water is ordinarily directly discharged without treatment when it is not recirculated. All fifty-eight plants included in the survey employ settling ponds for their lime cake slurry waste. Greater variation is found in disposal practice for waste water from flumes and beet washers, thirty-five plants employing ponding for various durations and twenty-three discharging directly to streams. On the average, about 32 per cent of the total volume of plant waste waters is discharged untreated. The remainder is usually screened (in forty-four plants) and at least a portion ponded. In a few plants, all wastes are ponded for short or long periods.

Data on waste composition were incomplete, but some generalizations can be drawn. First, although there is a sizable *volume* of waste receiving no treatment (condenser water, for example) the percentage of potentially *polluting materials* receiving treatment by ponding and other methods, is usually quite high. In other words, the liquid wastes receiving treatment, such as lime cake slurry, have relatively low volume, but these are generally the waste streams which carry a very high proportion of BOD and suspended solids. Typical BOD in the plant effluent before treatment (but after internal recirculation and recovery processes) ranges from 20,000 pounds per day to nearly 100,000 per day, with an "average" plant showing perhaps 40,000 pounds. After treatment, generally by ponding, discharge BOD appears to range from about 5,000 pounds to about 30,000, with an average at about 15,000. Figures for total solids, before treatment, range from 200,000 to about 800,000 pounds per day (of which 70 to 90 per cent are insoluble suspended matter), reduced after treatment to values between zero and 200,000 pounds per day. Suspended solids in raw wastes are in the range of 150,000 to 700,000 pounds per day; after treatment, generally by settling, suspended solids are discharged, usually at the rate of a few thousand pounds per day, but the quantities range from zero to 140,000, with one plant showing 200,000.[10]

On a basis of one ton of beets, the above BOD averages are about 15 pounds before treatment and 5 to 6 pounds after treatment; suspended solids are 50 to 200 pounds before treatment, zero to 50 after treatment. Based on the U.S. Public Health Service figure of 0.17 pound BOD per person per day, these wastes have BOD contents equal to those from about 90 people and 30 people respectively. A typical 2,700-ton plant, if not employing waste treatment thus discharges BOD in quantities equivalent to 240,000 persons, with variation from one-half to two and one-half times this figure. After treatment, usually by screening and ponding, final beet sugar plant wastes have BOD population equivalents commonly in the 30,000 to 175,000 range, with about 90,000 persons as typical. Suspended solids from a typical 2,700-ton plant employing settling ponds of a few acres capacity usually amount to about 3,000 to 10,000 pounds per day.

Table 5 shows the relative quantities of waste discharged—as pounds of BOD per ton of beets—from each processing step (taken from Table 3), from plants utilizing various process improvements, recirculation techniques, and waste treatment measures.

[10] The relationship between suspended solids and total solids in beet sugar plant wastes is usually such that either characteristic is a useful measure of waste content. The difference between the two is the dissolved solids, principally sugar and other soluble organic compounds. A small amount of soluble mineral salts (principally compounds of calcium, magnesium, sodium, and potassium) in the intake water and from beet processing is usually present also. The questionnaires show that the suspended solids comprise 70 to 90 per cent of the total solids in the combined untreated wastes from most beet sugar plants. The inorganic dissolved solids are not amenable to reduction or removal by any conventional treatment process, whereas the suspended solids may be mechanically removed by settling or filtration. Dissolved organic compounds may be at least partially removed by oxidation, the end products of which are mainly carbon dioxide and water.

Table 5

Representative BOD Quantities in Wastes from Plants Having Various Processes and Water Handling Procedures

(lb/ton beets processed)

	Potential generation	After process changes	After recirculation associated with process changes	After foregoing plus minimum waste treatment	After foregoing plus extensive waste treatment	After foregoing plus extensive recirculation and reuses
Flume water	4.5	4.5	4.5	4.5	2	1
Screen water	2.5[a]	2.5[b]	0	0	0	0
Press water	2.6	2.6[b]	0	0	0	0
Silo drainage	12.3	0[c]	0	0	0	0
Lime cake slurry	6.5	6.5	6.5	0.9	0.5	0
Condenser water	0.7	0.7	0.7	0.7	0.4	0.2
Steffens waste	10.4	0[d]	0	0	0	0
Total	39.5	16.8	11.7	6.1	2.9	1.2

[a] Average of BOD waste generation in batch diffusers and continuous diffusers.

[b] Assuming conversion to continuous diffusers but without recirculation of screen and press water.

[c] Assuming pulp-dryer installation.

[d] Assuming discontinuance of Steffens processing or the complete use of Steffens waste in by-product manufacture.

Most plants discharge their wastes to rivers, a few to lakes and estuaries, and two or three to municipal sewer systems. Four other plants are operating with essentially no waste discharge to surface waters, large ponds receiving and retaining the entire plant liquid wastes during the campaign. In these cases the volume of waste waters is greatly reduced through recirculation. Evaporation and seepage throughout the entire year eliminate all pollution, provided that the seepage does not contaminate any ground water aquifer. In a few plants, large ponds receive wastes during the campaign, little or no discharge to streams then occurring. Discharge from the retention basin in the season of high streamflow, usually in the spring, ameliorates the pollution effects.

The reported quantities of BOD and suspended solids discharged by beet sugar plants may be compared with statistics in earlier publications and with figures used in more recent government estimates. If "typical" figures of 15,000 pounds per day of BOD and 7,000 pounds of suspended solids are used to represent treated discharge from a 2,700-ton-per-day plant employing substantial waste treatment,[11] it may readily be seen that stream pollution is now much lower than the 60,000 to 100,000 pounds of BOD and 200,000 to 400,000 pounds of suspended solids applicable to general practice twenty years ago.[12]

[11] Assuming clarification of flume and lime sludge wastes.

[12] B. M. McDill, "Beet Sugar Industry," in *Industrial and Engineering Chemistry*, Vol. 39, No. 5 (1947), p. 657.

However, see U.S. Department of the Interior, South Platte River Basin Project, *The Beet Sugar Industry—The Water Pollution Problem and Status of Waste Abatement and Treatment* (June 1967). According to this government study, nine Colorado beet sugar plants in the South Platte River Basin, processing 24,000 tons of beets per day (1963–64 and 1964–65 campaigns), discharged to streams 334,000 pounds of BOD per day, an average of 14 pounds per ton of beets. This corresponds to 38,000 pounds per day from a 2,700-ton plant, more than double the "typical" figures from an improved plant indicated above. The BOD discharged by all but one of these plants covered a range from 7 to 19 pounds per ton of beets, and the final effluent from one plant (no longer operating) contained 34 pounds of BOD per ton of beets. Discharge of suspended solids was even greater—704,000 pounds, or 29 pounds per ton, equivalent to 80,000 pounds per day from a 2,700-ton plant. It is thus seen that in some localities and at certain periods in the processing campaign (measurements were made near the campaign end in most plants, when frozen beets result in higher waste loads), pollution may be much greater than national averages might indicate.

The current BOD load is equivalent to that of approximately 90,000 people rather than the 350,000 to 600,000 people previously estimated. It should also be observed that the seasonal operation of this industry limits the duration of discharge in most regions to about 100 days per year, but usually during periods of lowest streamflow. The total BOD discharged *per year* is therefore about one-fourth of that generated in a community of 90,000 people.

On a nationwide basis, a daily tonnage of about 160,000 tons of beets (assuming all plants to be operating simultaneously) yields about 800,000 pounds of BOD and 3 million pounds of suspended solids per day in the final effluents. The population equivalent of the BOD is about 5 million persons. Table 6 describes the overall BOD generation and discharge situation in the United States and also indicates the plant processes which give rise to the various waste streams. Table 6 shows figures for BOD generated and discharged in 1949 but it should be noted that the 1949 numbers are for only the fifty-six plants which were operating in both 1949 and 1962. Actually there were eighty-two plants in operation in 1949, with a total beet capacity of 150,000 tons per day, nearly as high as in 1962. Accordingly, the overall waste loads in 1949 were considerably higher than shown in the table.

Over the years, there have been rather large fluctuations in the sugar content of beets, primarily because of weather conditions, but there has been a general upward trend in this percentage. As a result, recoveries of sugar per ton of beets processed have also shown an upward trend. In addition, the introduction of continuous diffusers and the more extensive reuse of certain sugar-bearing water streams formerly wasted have permitted moderate increases in sugar yield. Partially offsetting these trends has been the decrease in Steffens processing, with the result that less sugar is produced and more of it is obtained in by-product molasses. The net effect of these changes on sugar yield per ton of beets, at least over the past fifteen to twenty years, has been small, and considerably less than that caused by seasonal variations.

These trends in increased sugar yield have been directly

TABLE 6

BOD in Beet Sugar Plant Wastes, 1949 and 1962[a]

(thousands of pounds per day)

Type of waste	1949: 113,000 tons beets per day					1962: 158,000 tons beets per day				
	BOD generated[b]	BOD removed by process changes[c]	BOD removed by waste treatment	Total BOD removal	BOD discharged	BOD generated[b]	BOD removed by process changes[c]	BOD removed by waste treatment	Total BOD removal	BOD discharged
Flume and washer water	510	x	x	100	410	710	x	x	270	440
Cooling water and condensate	80	x	x	10[d]	70[d]	110	x	x	30	80
Pulp screen and press water	550	50	70	120	430	840	630	60	690	150
Silo drainage	1,390[e]	660[f]	140	800	590	1,940[e]	1,920[f]	10	1,930	10
Lime cake slurry	730	0	350	350	380	1,030	0	960	960	70
Steffens filtrate	610	160[g]	80	240	370	770	560[g]	160	720	50
Total BOD	3,870			1,620	2,250	5,400			4,600	800

Note: x indicates that the quantities of BOD removed by process changes and by waste treatment are not individually available. The sums are known, however, and are shown as "Total BOD removal." The total BOD removed solely by process changes (the sums of the figures in the second column in each section of the table) would therefore be incomplete; but at least 22 per cent of the BOD generated in 1949, and 57 per cent in 1962, were removed by process changes.

[a] Figures based on production in the 58 plants operating in 1962, of which 56 were operating in 1949. Figures for 1949 apply to only 56 of the 82 plants operating that year, and cover only 113,000 of the 150,000 tons of daily processing capacity. If it were assumed that the remaining plants had water-handling methods equivalent to the 56 plants covered here, all BOD quantities in the 1949 portion of the table would be increased by $\frac{150{,}000-113{,}000}{113{,}000}$, or by a factor of ⅓.

[b] Based on BOD generated per ton of beets sliced in "unimproved" plant, as per U.S. Public Health Service, *Industrial Waste Guide* (reproduced in Table 3).

[c] By process changes, recirculation.

[d] Based on estimated 10 per cent reuse as diffuser makeup water.

[e] BOD which would be generated if all spent pulp were handled in silos (i.e., if no pulp driers were used).

[f] BOD not generated because of pulp drier use.

[g] By recycle and CSF production.

responsible for some decreases in waste production. The higher sugar content of beets has resulted in about the same waste generation per ton of beets handled and therefore a slightly reduced waste quantity per ton of sugar produced. Process improvements in the direction of higher sugar recoveries have had the effect of decreasing the losses of sugar to waste streams, per ton of beets and also per ton of product. And as previously explained, reductions in pure sugar recovery by discontinuance of Steffens operations has resulted in large absolute waste decreases by elimination of Steffens wastes and the substitution of salable by-products.

According to Henry,[13] European practice results in somewhat less BOD discharge than in the United States. At an annual beet tonnage of 100 million metric tons and 60 to 100 campaign days, 1.25 to 2.5 pounds of BOD per ton of beets (compared to the above estimated U.S. average of 5–6 pounds) would be in the final effluent for the population equivalent to be the stated 10 to 20 million persons.

In the U.S. example cited above, typical treated effluent was considered. For those plants employing little or no treatment of wastes and having less sophisticated waste recovery and water recirculation systems, BOD may be two to three times as high and suspended solids ten times as high as these figures. Thus, the figures formerly used for describing the wastes from beet sugar plants agree more nearly with practice in plants where there is very little water recirculation or waste treatment.

It may therefore be concluded that projection of these older figures into the future, without dealing with the process improvements and waste reduction practices that have been and are continually being established in the industry, will result in large overestimation of waste loads from the industry. In making such projections and, more broadly, in considering public policies for water quality management, it is very important that the potentials of waste water treatment and waste reduction through the recirculation of water streams in the plant be understood. The latter topic is the subject of the next section.

[13] J. M. Henry, *op.cit.*

Waste Reduction through Recirculation Several questions were included in the questionnaire pertaining to reductions in waste discharge which could be attributed to water recirculation in the plant. Usually, the respondents expressed the reduction in terms of pounds of BOD per day. It is clear that partial recirculation of water streams in the plant will reduce the *volume of water* discharged from the plant but will not, as such, reduce the amount of waste materials themselves unless some waste removal or recovery processes are used in the recirculation system. In some recirculation circuits, residual sugars are reintroduced into the production process and a portion of them winds up in the final product. Some of the respondents indicated that although a degree of recirculation was being employed, there was no BOD reduction in the discharged wastes. In a considerable number of plants, however, substantial reductions were indicated, due to the removal of wastes or to process changes which actually lowered the quantities of waste generated.

Illustrative of reductions due to waste removal associated with recirculation is the screening and settling of flume water and beet washer water prior to its recirculation. Like several treatment processes, this gives rise to a solid waste disposal problem. In the United States this is ordinarily not serious, because adequate space for disposal is available near the plants.

An example of waste reduction through recirculation that accompanies process changes is the elimination of pulp screen and press water waste by its total recirculation to continuous diffusers which have largely replaced the batch type. Recirculation is made practical by the adaptability of this new process to use of water which already contains suspended matter from previous drainage from extracted pulp. Nearly all of the pulp drainage water can thus be reused.

Quantities of BOD reduction achieved by these methods are shown in the questionnaires to vary from none to as much as 100,000 pounds of BOD per day. Common reductions of 10,000 to 25,000 pounds per day are claimed.

The wide differences are partly due to some respondents considering that waste reduction accompanying a process change, such as continuous diffusion, should be credited to the recirculation thus made practical. Other replies, involving the same process changes, did not directly attribute the waste reduction to the new recirculation arrangement. From previously reported waste concentrations, it may be deduced, however, that the elimination of pulp and press water through recirculation to the continuous diffuser eliminates about 5 pounds of the 30 to 40 pounds of BOD per ton of beets, or about 15,000 pounds of BOD in a typical 2,700-ton plant.

The total recirculation of flume and beet washer water, as practiced in at least two plants, effects almost as much BOD reduction as process water recirculation. This stream normally carries about 4 to 5 pounds of BOD per ton of beets, equivalent to about 12,000 pounds of BOD in a typical 2,700-ton plant. If condenser water is also recirculated completely, another 0.7 pound BOD can be eliminated, and the recirculation of the lime cake drainage water from the lime settling pond, rather than its discharge, eliminates an additional 0.9 pound.

Thus, recirculation and attendant treatment or materials recovery itself, exclusive of such waste reduction practices as the ponding of lime sludge, the ponding or processing of Steffens waste, and the elimination of silo drainage by pulp drying, can make a maximum reduction of about 11 to 13 pounds of BOD per ton of beets, or about 30,000 pounds in a 2,700-ton plant. This is a population equivalent of about 175,000 or about 35 per cent of the potential waste load (as defined in Chapter 3) from the plant. Combining maximum recirculation with harmless disposal of Steffens waste, ponding of lime sludge, and elimination of silo drainage can reduce the waste load coming from the plant nearly to zero.

If there is complete recirculation of these four streams (flume and washer, pulp screen and press, lime overflow and condenser) elimination of suspended solids by settling could average about 60 pounds per ton, or 160,000 pounds in a 2,700-ton plant, depending on the highly variable amount of soil adhering to the beets.

In these appraisals of the effects of maximum recirculation, the removal of the ninety pounds or so of solid lime sludge by settling is not included as a recirculation benefit because it is done in all U.S. plants whether or not there is any recirculation of the decanted solution.

Waste Reduction per Ton of Beets Table 5 shows, per ton of beets, the BOD discharged from the various operations in sugar plants employing several processing and waste handling methods. The figures may be considered typical of each type of plant, subject to the assumptions noted in the table. Where zero discharge is indicated, it is of course assumed that either a waste has not been produced at all, due to a process change, or that total recirculation or complete waste treatment has been provided. The last three columns in the table contain figures representing progressively increasing degrees of waste treatment, and the quantities shown are therefore somewhat arbitrary and dependent on the actual extent of treatment. The few plants from which there is no surface effluent would obviously involve zero BOD discharge, a situation not fully reflected even in the last column of the table.

It is seen that pulp drying and the elimination of Steffens waste reduce BOD discharge to about 40 per cent of the potential generation and that the relatively simple recirculation of process water to continuous diffusers makes an additional reduction to about 30 per cent. Settling of lime slurry cuts the remaining BOD discharge in half, to about six pounds per ton of beets, or 15 per cent of potential generation. This level may be considered that which can be achieved by comparatively simple and economical methods. Further waste reduction requires extensive treatment or recirculation systems, primarily for the handling of flume water.

Current Practice and Trends in Water Recirculation and Waste Treatment Processes

Technological change in beet sugar manufacture, particularly during the past twenty years, is directly and indirectly reducing the industry's intake of fresh water, both in terms of water per ton of beets and in terms of the withdrawal by the entire industry. This, in turn, reduces the quantities of waste waters discharged and, in this case, their waste content. In many instances the reduction in water intake has, in fact, been the result of facilities primarily meant to control waste water streams. The changes and their effects are discussed in the subsections below. The first six sections deal with process changes which affect the amount and character of the plant effluent before treatment. The remaining sections deal with waste water treatment processes. Questions were asked in the supplementary questionnaire concerning the dates of introduction of various processes for water conservation and waste reduction and waste treatment processes. In general, the replies show that there has been a gradual but fairly steady introduction of practices along all three lines. The questionnaire results are summarized in Table 7.

TABLE 7
Recirculation and Waste Treatment Processes in Beet Sugar Plants, by Time of Introduction

	Number of plants					
	Before 1910	1910–30	1930–50	1950–60	1960–62	Total[a]
Water recirculation and reuse (54 plants):						
Recirculation of flume and washer water	–	2	11	4	3	23
Recirculation of pulp screen and press water	–	1	4	11	25	41
Cooling water reuse in flumes and washer	4	8	8	3	1	25
Cooling water recirculation for cooling	–	1	8	2	3	14
Silo drainage elimination by pulp drying (58 plants):	–	16	8	12	8	49
Waste treatment processes (58 plants):						
Screening	7	7	5	14	7	44
Sedimentation[b]	–	4	14	10	10	45
Biological filtration	–	–	–	–	1	2
Evaporation (other than in ponds)	–	–	2	1	1	6
Stabilization ponds[b]	1	3	12	13	4	38
Regulated discharge[b]	–	–	4	4	1	13

[a] Rows usually do not add up to totals because some plants did not report when an operating process was initiated.

[b] Several waste treatment processes involve ponding of effluents. In some cases only one process occurs in a waste pond, i.e., sedimentation. In larger ponds, both sedimentation and stabilization may occur. And if regulated discharge is practiced, the pond performs all three functions. Evaporation (from ponds) is an incidental, but significant, factor also, but the tabulated figures do not include this natural process.

Continuous Diffusers Only a comparatively few plants (sixteen out of fifty-eight) had not yet converted from cell-type to continuous diffusers in 1962. Most of the conversions (twenty-eight) have occurred since 1949. Several of the sixteen plants have plans for conversion within the next few years. In addition to direct process economies, primarily due to labor saving, the adoption of continuous diffusers usually results in a decrease in both gross water use and in waste production. The reduced water use and the inherent characteristics of a continuous process reduce the cost of recycling essentially all of the pulp screen water and pulp press water to the diffuser. Thus, these two sources of pollution have been greatly reduced in the industry. The net result of this change is a decrease of about 4.6 pounds of BOD per ton of beets processed by the new method, a total decrease approaching 600,000 pounds of BOD per day from the forty-two plants employing continuous diffusers.[14] The population equivalent of the BOD figure is about 3.5 million. When the remaining plants have converted to the new process, there should be a further decrease of about 125,000 to 150,000 pounds of BOD per day. With approximately 18.5 million tons of beets processed per year, annual BOD reduction by use of continuous diffusers and recirculation of screen and press water should presently approach about 85 million pounds.

The above figures must be regarded as maximum theoretical values which have not been fully realized. One cannot assume production of 4.6 pounds of BOD in all plants which formerly used cell-type diffusers; there were plants employing partial recirculation of pulp screen and press waters.[15] Hence, the waste generation reductions due to conversion to continuous diffusers are less than the above estimates, possibly by one-fourth or one-fifth. It may therefore be estimated that the adoption of

[14] Figures on waste contained in particular process effluents published by the U.S. Public Health Service (Table 3) are used in making these calculations. See *Industrial Waste Guide to the Beet Sugar Industry*, U.S. Public Health Service, 1950.

[15] Two plants still using cell-type diffusers in 1962 were recirculating all their screen and press water. Moreover, seven of the plants employing the newer continuous diffusers did not recycle the screen and press waters in 1962, instead simply ponding them or otherwise disposing of them as formerly.

continuous diffusers and the accompanying water recirculation practice have resulted in a daily BOD reduction of roughly one-half million pounds during the processing campaign. Of this reduction, nine-tenths is estimated to have occurred since 1949.

Pulp Drying One of the most noxious wastes of the beet sugar industry, one with exceptionally high BOD content, has been drainage from silos in which wet beet pulp is stored prior to sale as cattle feed. This waste contains about 12.3 pounds of BOD per ton of beets processed. If the pulp can be sold immediately, no storage is needed and there is no silo drainage. However, the uncertainty of immediate demand and means of transport has usually dictated the use of storage silos for practically all of the wet beet pulp.

Prior to 1949, twenty-three plants had eliminated this waste by installing pulp drying equipment, and subsequent to 1949, twenty-one additional plants have adopted drying. Five other plants produce dried pulp (year of process installation was not reported). In 1963, only nine plants had not yet installed pulp dryers. It may therefore be concluded that silo drainage no longer constitutes an important source of waste from this industry. Compared with the waste load that would be involved if all plants were using pulp silos, the BOD reduction is about 12 pounds per ton of beets, or 1.92 million pounds per day, corresponding to 225 million pounds per year and a population equivalent (during campaigns) of 11 million persons. It may be estimated that actual BOD reduction since 1949 due to adoption of pulp drying has been about one million pounds per day.

There may be an economic advantage, completely aside from pollution abatement, in converting to pulp drying, because if there is a nearby market for dry pulp, this operation yields a profit to the plant greater than does the sale of wet pulp. The rate of conversion to the dry product has depended mainly on the establishment of a dependable market. This, in turn, involves transportation considerations, long distance shipping generally not being economical. Conversion of the remaining plants can be expected to continue at a reasonably steady pace (one company executive indicates dryer installations at the rate of about one per year). It is reasonable to anticipate that within five to ten years, practically all the beet pulp produced in the industry will be sold as a dried product.

Steffens Waste Processing Of the fifty-eight plants operating in 1962, twenty-five were using the Steffens process in 1949. Two of these plants discontinued Steffens operations in 1954 and one in 1962. Of the twenty-two Steffens plants operating at that time, fourteen have installed evaporation equipment for concentrating the waste prior to mixing with beet pulp or shipment to another plant for extraction of by-products. Since this further processing is accomplished with very little waste discharge, Steffens waste thus remains a significant factor in only eight plants in the United States. These eight plants handle about 18,400 tons of beets per day (about 12 per cent of the industry) and approximately 900 tons of molasses.[16] At an estimated 231 pounds of BOD per ton of molasses, the Steffens filtrate produced in these plants contains about 210,000 pounds of BOD per day.

Six of these eight plants treat Steffens waste by ponding. There is no discharge from five of these Steffens waste ponds, and the sixth effects some BOD and suspended solids reduction between the end of the campaign and spring release to the stream. Four of these ponding systems have been installed within the last ten years. Final effluent figures are not available, but it may be reasonably well estimated that the release of Steffens waste to streams is now contributing only about 70,000 pounds of BOD per day. Although an occasional local situation may be importantly affected by this particular waste stream, on a national basis pollution from this waste has been substantially reduced.

Since 1943, the fourteen plants which have installed concentrators for Steffens waste have eliminated approximately 550,000 pounds of BOD per day. Of this amount, about 320,000 pounds have been since 1949 (from nine plants). Several sugar plants ship concentrated Steffens filtrate (CSF) to a single fac-

[16] A minor portion of this molasses tonnage is by transfer from non-Steffens to Steffens plants.

tory for monosodium glutamate production. Others mix and dry CSF with spent pulp to enhance its protein content for animal feeds. Another five plants have simply discontinued Steffens operations and sell the molasses or mix it with pulp for feeds, thereby eliminating about 120,000 pounds of BOD per day, 75,000 pounds of which have been since 1949.

The molasses from numerous beet sugar plants not using the Steffens process is handled in several ways. It has considerable value (averaging $30 to $40 per ton) due mainly to its sugar content. Disposal of this by-product may be by shipment to a Steffens plant where it is mixed with molasses produced from beets processed at that location. In some factories the molasses is mixed with dried beet pulp to yield an animal feed of higher carbohydrate content. The remaining non-Steffens plants sell molasses for the manufacture of fermentation alcohol and for mixed livestock feeds.

Although the market for monosodium glutamate, the principal by-product from Steffens waste, appears reasonably stable at the present time, there is reason to expect a gradual growth in the use of this product. However, a new fermentation process for producing monosodium glutamate, from an entirely different raw material, has been introduced, so the costs of these processes will affect the demand for CSF. It appears unlikely that there will be further expansion in the use of CSF for monosodium glutamate production. But the expansion in use of CSF as an additive to dried pulp is a factor which makes it reasonable to expect a gradual adoption of Steffens waste concentration in the several Steffens plants not yet using the concentration process.

It is not considered likely that Steffens waste will entirely disappear in beet sugar plant effluent streams, through its further processing for by-products. It is probable, however, that within the next five to ten years this waste will be reduced to a negligible quantity. Final effluents will, of course, depend on ponding and other waste handling methods. Stringent pollution control policies could, however, contribute to even greater CSF and by-product production as the cheapest of the several alternatives.

Recirculation of Condenser Water In contrast with the process changes outlined above, the primary purposes of which have been manufacturing economies and by-product production, this and the following changes are directed specifically at the reduction in intake water requirements and the decrease in waste production in beet sugar plants. The recirculation of condenser water, through cooling towers or spray ponds, is very common in the chemical, petroleum refining, and electric power industries, but it has not been widely used in beet sugar plants. Only thirteen plants had adopted this procedure by 1962. Since that year, however, several additional recirculation systems have been installed. Condenser water has a low BOD concentration (about 40 ppm) but because of its large volume, it is of considerable importance in the remaining discharges of the industry. Also, the temperature increase in this stream is substantial. Consequently, major pollution control advantages, as well as reduction in water withdrawals, can result from recirculation. Where raw water is in limited supply, or where pumping costs (as from wells) are high, or where excessive sediment or other materials may require removal prior to water use, the most economical solution is frequently the recirculation of condenser water through cooling towers or spray ponds. In humid climates, cooling towers are usually employed, whereas in dryer regions spray ponds or cooling lagoons may be used. Only the makeup water, equivalent to as little as 10 per cent of the circulating stream, need then be withdrawn from the supply source. A few plants in which this technique is used have greatly reduced water withdrawal. For example, some plants employing this practice have reduced total water intake to less than 1,000 gallons per ton of beets, compared to the industry average of about 2,600 gallons per ton.

In a recent study by Cootner and Löf[17] the costs of water for power plant cooling have been examined and relationships for projecting the cost and use of water have been developed. If cooling towers are used in the recirculation system, the total of capital, power, and treatment costs usually ranges from one to

[17] Paul H. Cootner and George O. G. Löf, *Water Demand for Steam Electric Generation* (Washington: Resources for the Future, Inc., 1965).

five cents per thousand gallons, depending on size of installation, atmospheric conditions, and other factors. If fresh water costs exceed these levels, or if the warmed condenser water is not to be used for other purposes, such as beet washing, recycling through cooling towers or ponds would be desirable. Similarly, pollution control considerations may make these procedures desirable.

It is interesting to observe that one California plant cools the used condenser water not for recirculation but for thermal pollution abatement. A cooling tower is used for the condenser water effluent, thus avoiding temperature rise in the receiving stream. The net cost of this operation appears to be in the range of 1.5 cent per thousand gallons, roughly equivalent in this plant to 1.5 cent per ton of beets or 10 cents per ton of sugar produced.

When atmospheric conditions are not hot and humid, an open pond can provide the necessary cooling, if properly sized. In all states except California, the processing campaign is limited to the fall and winter seasons, when weather conditions make cooling ponds practical. However, most of the plants employing condenser water recirculation are in California, where cooling towers are usually more economical than ponds.

Although condenser water has the lowest BOD concentration of beet sugar plant wastes, its pollution effects are by no means negligible. The BOD discharged in this stream appears to be about 75,000 pounds per day in the United States. This quantity is larger than the present BOD discharged in Steffens wastes or in lime pond overflows, and several times larger than in silo drainage. The high dilution results in a major disposal problem, however. More complete recirculation, with possible in-plant reduction by aeration, may lead to a solution.

There have been a few experimental attempts in other industries to employ cooling towers also for BOD removal, on the principle of the trickling filter. By use of tower packings of various plastic media, organic growths may become established on these surfaces and serve to decompose (oxidize) various dissolved organic compounds. Partial and limited success has been achieved in cooling and oxidizing oil refinery wastes containing dissolved phenol, as well as paperboard mill wastes. The principal problems are associated with excessive fouling of the tower packing and destruction of micro-organisms by temperatures higher than their tolerance. There appear to have been no attempts to try this technique with beet sugar condenser water, but the possibility seems to exist.

Recirculation of condenser water for repeated use may be expected to increase, but it is in competition with another type of reuse, outlined below. Unless raw water costs (including any treatment and pumping expenses) rise above about one cent in some climates to as high as three cents per thousand gallons, condenser water recirculation through cooling towers will probably grow at only a slow rate. Thus, water for condensing will probably continue to be a fairly large part of intake in the beet sugar industry. This trend could be affected, however, by possible restrictions on the discharge of heat or BOD.

Condenser Water Reuse for Flumes and Beet Washers Only four plants have adopted condenser water reuse for flumes and beet washers since 1949, but twenty-one had used the process prior to that time. Thus, about one-half of the plants in the industry follow this practice and almost all of these have for some time. There is a decided "company pattern" to this and other recirculation practices, most of the plants in a single company following the same procedure. Where plants are not recirculating all the used condenser water for further cooling, and where water conservation is at all important, a simple and economical procedure is the reuse of condenser water for conveying and washing beets. Benefit is also realized from the heat in the water in plants located where beets become frozen in outdoor storage piles.

The quantities of water used for conveying and washing beets are usually about half of the total gross use in the plant. The average in forty-three factories is 2,230 gallons per ton of beets. This water may be entirely fresh water, or it may be used condenser water, recirculated flume and washer water, or various combinations of these and other streams. Usage varies widely. One plant needs only 525 gallons (per ton of beets) for this

3,000 gallons per ton of beets processed. In most plants (seven-purpose, ten employ less than 1,600 gallons, and seven use over teen) in which the warmed water is used for conveying and washing beets, the major part of the fluming and washing water (70 to 95 per cent of it) has previously been used for condensing. In the remaining twelve plants for which recirculation rates were reported, the proportion of condenser water in the total flume-washer water covers a range from 11 per cent to 63 per cent. Average percentage of condenser water in the flume-washer streams in twenty-nine plants is 71 per cent, the balance usually being either fresh water or recirculated flume-washer water. (A few plants use part or all of the pulp screen and press water or a mixture of all plant wastes for fluming beets.)

In most plants employing this type of water reuse, practically all of the used condenser water is supplied to the fluming and washing system. Seventeen plants recirculate over 90 per cent of the condenser water in this manner, another eight plants reuse from 50 to 90 per cent, and four plants reuse less than 30 per cent for fluming and washing beets. Of this last group, three plants recool the major portion of the water for recirculation to further condensing use.

The small concentration of sugar in the used condenser water occasionally may be a problem due to a tendency to fermentation. This can be controlled by chlorination, if necessary, at relatively small cost.

There is reason to expect that the industry will make moderate additional use of this water-saving technique in areas where demands for reducing water intake become pressing.

Recirculation of Flume Water and Beet-Washer Water About one-third of the U.S. plants practice some recirculation of the water used for conveying and washing beets. Most of these plants have followed the practice for more than fifteen years, but at least three plants have adopted this arrangement during the past five years. In most of these plants, water leaving the flume and beet washer carries large amounts of sediment, beet fragments, leaves, and other debris. These materials are usually removed by screening and sedimentation (in settling ponds or mechanical thickeners) prior to recirculation. Chlorination may also be employed to reduce fermentation of the dissolved sugars present. Practice varies from recirculating the stream to only a limited extent (equivalent perhaps to one reuse), to the other extreme where this water is almost totally recirculated, through a large retention basin.

From figures previously cited, sampled from questionnaires related to pond operation, it may be inferred that in those plants in which water from flumes and beet washers is recirculated and stored in ponds of one to three days' capacity, suspended solids can be reduced 20 to 70 per cent and BOD 10 to 30 per cent from the levels that would exist if no recirculation and waste ponding were practiced. In other words, if once-through use of water for fluming and washing beets is practiced, without ponding or other effluent treatment of any kind, the wastes entering streams would be larger than now prevail in plants using ponding and recirculation, by the factors estimated above. As previously described, several plants (four) practice almost total recirculation, so elimination of BOD and suspended solids from flume and beet washing water in these plants approaches 100 per cent.

The most common practice among twenty-six plants employing recirculation of water from the fluming and beet washing operations is the return of water in eighteen plants to flumes without treatment (other than grit removal in several plants). Ponding (primarily for solids settling) prior to recirculation is practiced in six plants, and mechanical clarifiers are used for this purpose in two plants. In most of the plants recirculating flume water without treatment, the *final waste water* from these operations is ponded as indicated below. It may thus be concluded that presently on an industry-wide basis, *recirculation* of flume- and beet-washer water has only a small direct effect on the reduction of waste load discharged and, hence, on stream pollution, and that the principal factor is the *ponding of the final effluent,* regardless of recirculation practice.

Rather than being particularly important as a pollution control measure, recirculation of flume and beet washer water, as presently practiced, is a significant factor in the reduction of water intake and in the total *volume* of final effluent. The weighted average total fresh water withdrawal in twenty-two of the plants practicing this type of recirculation is only 1,650 gallons per ton of beets, whereas the average is 2,900 in all other plants. This saving of 1,250 gallons per ton, largely due to flume water recirculation in these twenty-two plants, represents 85 million gallons per day for this segment of the industry. The daily water saving in all twenty-six plants recirculating this stream is about 100 million gallons.

In addition to water withdrawal savings, there are, of course, reductions in waste discharge volume. Indirectly, these factors also result in modest reduction in BOD and suspended solids discharge in most plants because the settling ponds and stabilization lagoons for the final flume waste water perform more effectively than in plants not recirculating this stream. The roughly 40 per cent average reduction in effluent volume from plants employing flume water recirculation permits a corresponding two-thirds increase in residence time for wastes in ponds of fixed capacity; a higher degree of removal of BOD and suspended solids thus occurs. In the half-dozen plants using a very high recirculation rate, such that intake is reduced to less than 1,000 gallons per ton, ponds of a given size have about the same waste removal effectiveness as ponds three times as large in plants employing no recirculation.

Simultaneous Recirculation of Flume-Washer and Condenser Streams If found desirable, it is possible at some cost (to be discussed in detail in Chapter 5) to simultaneously recirculate both the condenser- and flume-washer waters in a closed cycle. This may be accomplished either by recirculating each one back to the process from which it came or by interconnecting both in a common recirculating system. The latter type of system is shown in Figure 5, Chapter 3.

In either case, it is necessary to have intervening treatment processes (ponds or others) which cool the condenser water, remove solids from the flume-washer waters, and keep degradation processes under control in order to avoid fouling of equipment, air pollution, and other problems. This procedure can virtually eliminate external discharge of any wastes originating in the condensing, fluming, or washing processes.

Waste Treatment Practices Turning now to waste treatment practices and the time of their adoption, the questionnaires show that the most commonly used method is screening, followed closely by settling and sedimentation. Nearly forty plants report the use of screening, and of these, twenty have installed the equipment since 1949. Sedimentation is reported in use by thirty-seven plants, eighteen of which employed the process prior to 1949. But since response to the specific questions on settling ponds indicated that all plants employed at least a lime settling pond, it must be concluded that there were some differences in interpretation of the sedimentation question and that all plants actually do use this process at least to some degree.

Although the questionnaires indicated that only four plants used evaporation as a waste treatment method,[18] the question was probably misunderstood because fourteen plants reported Steffens waste evaporation to produce CSF. It may be inferred that the processing of waste by evaporation is limited to these fourteen CSF producers.

All plants reported the use of ponding for part or all of the plant waste. In most instances, these reports correspond with those for sedimentation, the two processes occurring in conjunction with each other. Of the plants indicating when retention and sedimentation ponds were first used, four show adoption since 1960, thirteen between 1949 and 1960, and twenty-one prior to 1949. Twenty questionnaires did not indicate dates.

As to types of wastes treated by ponding, every U.S. plant reported the retention of lime cake slurry. Provided that the pond is of adequate size, the contamination of surface streams by suspended lime (actually calcium carbonate or chalk) is effectively eliminated. Unless the pond is very large there is some overflow of the settled water—generally to surface streams—and soluble wastes, including BOD, are only partially re-

[18] This refers to processes where evaporation is a central objective. All ponding, even for sedimentation, is accompanied by some evaporation.

moved. About half of the plants have no overflow from the lime ponds. Eight plants discharge to streams 8 to 12 per cent of their lime sludge pond inflow, and one plant discharges overflow to municipal sewer works. The average lime sludge generation in nineteen plants is 87 gallons per ton of beets.[19] Using figures of 6.5 pounds BOD and 90 pounds of suspended solids per ton of beets, the thirty plants having lime sludge ponds with little or no discharge remove one-half million pounds of BOD and 7 million pounds of suspended solids per day. For the remaining twenty-eight plants, we estimate an average discharge of one-third the inflow, that is, about 30 gallons per ton of beets, carrying dissolved BOD equivalent to about 0.9 pound per ton of beets and essentially no suspended solids. These figures correspond to the removal of 465,000 pounds of BOD and another 7 million pounds of suspended solids. The removal of these two pollutants thus totals about 93 per cent and practically 100 per cent, respectively, of which over half has occurred since 1949. The remaining 65,000 pounds of BOD in the lime pond effluent to streams is equivalent to a population of about 400,000.

Six plants reported the ponding of Steffens wastes, and twenty plants show the ponding of flume- and beet-washer water In two plants, a portion of the flume water is ponded. The ponding of flume water and beet-washer water (whether or not recirculation may be used) was estimated from the questionnaire results to accomplish a BOD reduction, mainly through ponding of the waste water, but also by screening and partial or total recirculation, of about 270,000 pounds per day. This is an average reduction of about 40 per cent. These estimates were derived from reports of treatment practices, pond sizes, quantities or composition of inflow and outflow, and pond retention periods.

Seven plants indicated the use of ponds for condenser water, and six show the ponding of silo drainage and pulp screen and pulp press water. Six plants stated that "general wastes," not otherwise specified, are ponded, and eight plants (including

[19] This figure agrees closely with the long-used 90-gallon estimate quoted in Table 3.

some of the above) showed the handling of all plant wastes of all types in individual or separate ponds.

The principal function of most ponds is their use for settling suspended solids, primarily precipitated lime cake (calcium carbonate). If only the lime cake slurry is ponded, the removal of this suspended matter ranges from about 100,000 pounds to 200,000 pounds per day, removal in the average plant being approximately 150,000. Roughly, then, the retention of lime wastes by the fifty-eight U.S. beet sugar plants operating in 1962 prevented about 14 million pounds of suspended matter, per operating day, from entering surface waters.

Practice varies considerably in the separation of wastes into various ponds. In plants where ponds are used for more than lime wastes, the lime pond is usually separate from all others. In some instances, a pond may serve as temporary storage for a recirculated "waste," such as condenser water. In these cases, the wastes are usually not mixed, individual streams being maintained separate for recirculation use. Steffens wastes are handled separately or are sometimes ponded with the lime waste. Flume waste is handled in ponds which are considerably larger than those for lime and Steffens waste, in order that the retention time can be sufficient for most of the suspended matter to settle.

Although ponds are effective in the removal of suspended solids, they do not usually have much effect on soluble BOD. Short-term retention (one to three days) results in the removal of BOD associated primarily with the insoluble matter which settles; during the same time, dissolved BOD, principally as sugars, is only slightly oxidized. In most plants ponding only lime waste, BOD removals of 10,000 to 20,000 pounds per day are typical, with perhaps 15,000 as an average. When flume water is also ponded, the combined BOD removal is commonly in the 25,000 to 50,000 pounds per day range. These figures are typical of those plants where ponding provides retention of one to three days' time.

The removal of suspended solids from wastes other than lime cake appears important in many plants. Ponding of flume- and beet-washer water, which is responsible for the largest suspended solids load, typically results in the removal of 200,000 to as much as one million pounds of solids per day, with an average in the range of 400,000 to 500,000 pounds. This figure is highly variable depending upon the plant size and the cleanliness of the beets when they are received. Much of the solid material carried by the flume water is soil particles washed from the beets; soil moisture content during the beet harvest materially affects the solids content of the flume and wash water.

A large variation was noted in pond size and the resulting waste retention time. The period varies from as little as one-half day to as much as thirty days in those plants which can be considered to have steady-flow type ponding. Even the shortest of these retention times appears to be adequate for eliminating the major part of the suspended solids from lime wastes and flume waters, and some BOD removal by sedimentation also occurs.

Nineteen plants reported that there was actually no surface discharge from the waste ponds they operated. Ten of these plants used ponds for lime waste only or for a combination of lime and Steffens wastes; flow rates to the ponds ranged from about 100,000 gallons per day of lime cake slurry up to something over one million gallons per day of mixed lime and Steffens wastes. The ponds are large enough to retain all the wastes of these particular types for the entire campaign and to eliminate the water by seepage and evaporation during the campaign and through the balance of the year. In one plant, for example, with a daily beet throughput of 4,300 tons, 1.24 million gallons of Steffens and lime wastes enter the pond where 450,000 pounds of suspended solids are removed per day. There is no surface discharge, seepage and evaporation thereby accounting for a total annual dissipation of about 120 million gallons of water. In a 2,000-ton plant, a daily flow of 125,000 gallons of lime cake slurry enters the pond where 135,000 pounds of solids settle per day without surface outflow.

In other plants reporting no outflow from ponds, wastes in addition to lime and Steffens are included. Four plants on the West Coast retain all the plant waste waters in large lagoons where total disposal is by settling, seepage, and evaporation throughout the year. (Odors resulting from anaerobic decomposition may be an occasional nuisance.)

In most plants, where one to five days' pond retention time is provided, the pond overflow is only slightly less than the inflow, the difference being seepage and evaporation. Typical reductions in volume of 10 to 20 per cent occur during these short retention periods. As previously indicated, BOD reductions are almost entirely due to settling of suspended matter.

A third type of pond operation features retention times of over 100 days (approximately the length of the processing campaign) and the regulated discharge of wastes during and following the campaign. Seven plants, all in the Midwest, operate on this basis. Retention times vary from about 120 days to 200 days. A moderate pond overflow rate is usually maintained during the campaign, but in a few plants there is no discharge during this period of low streamflow. Most of the wastes are discharged the following spring when rivers are at high level and ample dilution is provided. The schedule is regulated by consideration of river freezing, streamflow, and pond capacity.

In one plant, wastes are ponded for subsequent primary and secondary treatment in the municipal sewage system. Controlled flow to the city sewer is started after the campaign, when decomposition has reduced the BOD content of the effluent, the period of discharge usually being about one month. Release is usually limited to night hours, thereby balancing the load on the municipal sewage treatment plant. Charges assessed by the city for this service include $24.00 per hundred pounds of BOD in excess of 200 parts per million and $18.75 per hundred pounds of suspended solids in excess of 240 parts per million. A plant in Idaho will adopt this same practice in the near future.

Summarizing the ponding of beet sugar factory wastes, we see that all plants treat some of the waste by this method, and a few plants treat all of the wastes by ponding. Most plants employ short-term ponding and discharge an overflow after partial elimination of suspended solids and BOD. In a few factories, there is no surface discharge from ponds treating small volumes of highly loaded wastes, and about a half-dozen plants have no surface discharge from large lagoons handling all of the plant wastes. Regulated discharge from large ponds in about ten plants in the Midwest permits release during periods when the wastes contribute the minimum nuisance to surface streams.

These considerations demonstrate that ponding is already a partial solution and can be a *total* solution to the water pollution problem in the beet sugar industry.[20] The primary difficulty is the large size of ponds necessary for complete waste elimination, hundreds of acres being required by some plants. The availability of land at reasonable cost is therefore a prime factor in this method of handling the problem. Another difficulty is the nuisance created by odors from open ponds containing decomposing organic materials. In some locations, this air pollution problem might outweigh the water pollution difficulty. Progress in actual waste reduction by process changes is, however, reducing the volume requirements for retention ponds. Recirculation of various water streams further reduces the waste volume and hence the necessary land area. These several industry changes are leading to the possibilities for total waste elimination through the ponding process.

New Waste Treatment Processes It is apparent from the trends in the beet sugar industry that the several wastes and sources of pollution are gradually decreasing to a single process waste, the overflow water from conveying and washing beets.

The handling of this large volume of waste, containing high loadings of BOD and suspended solids, has posed the largest disposal problem to the industry. Because of the volume involved, ponding is expensive, and not universally practical. Since this stream has become the primary waste problem of the industry, considerable effort is being devoted to the development of new processes for handling it. One of these processes employs the trickling filter, sometimes used in municipal sewage treatment plants. Limited success has been achieved with this method, on

[20] If conditions are such that seepage results in pollution of nearby ground water supplies, ponds may have to be provided with water-tight linings; water disposal must then be entirely by natural evaporation. This extreme measure would not usually be required.

an experimental, small-scale basis. Another technique which has been tried is broad field disposal, by which process the waste is allowed to flow slowly over a large, slightly sloping field in which grass and other plants are growing. Settling of solids and some oxidation of BOD occur, with an improvement in the quality of the waste stream. This method encounters some of the same limitations as does the ponding process, such as large area requirements and the need for occasional changes in the land area used.

The Beet Sugar Development Foundation is supporting a study in the British Columbia Research Council for the development of a biological oxidation process for handling flume- and beet-washer wastes. It has been found in preliminary experiments that some strains of soil bacteria naturally occurring in the wastes can greatly reduce the BOD content. If this process proves successful, ponding of the waste would be involved, but probably with shorter retention times and more complete removal of pollutants than now are possible with simple ponding.

Still another approach is by the biological treatment (activated sludge process) used in municipal sewage systems; one plant is actually employing this method now for waste disposal through the municipal plant. Further development of this technique specifically along lines best suited to the constituents of this beet sugar plant waste may be a practical solution in some cases. The problem is complicated, however, by start-up difficulties due to the intermittency and relatively short duration of the beet processing campaign.

Another experimental waste treatment system comprises a series of three deep ponds of decreasing depth, through which the plant effluent flows successively. Anaerobic decomposition takes place in the first pond, a transition to aerobic conditions occurs in the second, and aerobic oxidation prevails in the third pond. Pilot plant experiments are in progress.

These treatment processes are in addition to a recirculation system previously described and actually used in one plant for several years. This is complete recirculation of flume and washing water. In this particular plant, flume, wash, and condenser waters are combined in a single recirculation system. Although odors are occasionally encountered, the practice is generally satisfactory and improved versions, in particular, would be amenable to use in other plants.

One of the most promising developments in the handling of flume and beet washing water is the use of a series of small, deep ponds through which the screened water is circulated. Addition of lime, in the form of a calcium hydroxide suspension, immediately prior to screening and ponding materially aids the settling of solids and virtually eliminates anaerobic decomposition. The first pond is in the form of a U-shaped channel, about 8 feet deep, 20 feet wide, and 1,200 feet total length. Most of the solids settle in this pond as the water flows through its length, with a retention time of only about 4 hours. Removal of sediment after the campaign (and during the campaign if necessary) is easily accomplished by use of a crawler-type dragline excavator on the channel bank. After passage through the U-shaped channel, the partially clarified water flows slowly through a series of two small ponds, positioned "inside the U," for completion of settling. The total area of the five ponds is only 2 acres, the total volume approximately 15 acre-feet (5 million gallons), and the retention time about 14 hours.

Operation of this new system during the 1966–67 campaign at one plant was comparatively free of problems. Total recirculation was employed throughout the entire campaign, with only occasional small overflow to a waste pond from which there was no discharge. Odor was not excessive, and water returned to the plant was of satisfactory quality for beet conveying and washing. Average removal of about 85 per cent of suspended solids was achieved, while BOD gradually increased to about 3,000 ppm, where it remained approximately constant during the last six weeks of the campaign.[21]

The total investment in this treatment facility, comprising vibrating screens, lime-feeding equipment, ponds, pumps, and piping is about $165,000 for this 2,200-ton-per-day plant. Its successful use has led to a decision to install a similar system in

[21] Lloyd Jensen, Vice President, The Great Western Sugar Company, personal communication, January 1967.

a nearby 3,400-ton-per-day plant, at an estimated cost of $230,-000. It appears that an investment of about $75 per ton of daily beet capacity is required for these facilities.

Lime has been found to be an important factor in the effectiveness of this system, both for achieving satisfactory settling and also for avoiding septic conditions in the ponds. A pH of about 8 (in the water returned to the fluming operation) appears near ideal, requiring the addition of 3 pounds of lime per ton of beets processed.[22] At a lime cost of 0.75 cent per pound,[23] other operating costs of 0.63 cent per thousand gallons circulated,[24] and annual capital costs of 10 per cent of investment, total costs of recirculation and accompanying pollution abatement are 12 cents per ton of beets processed, roughly equivalent to one dollar per ton of sugar produced. As indicated below, this expense is considerably less than the costs of other complete waste removal processes. If further development and testing substantiate these first favorable indications, the problem of economical abatement of pollution from flume- and beet-washer wastes may be reaching a solution.

A similar treatment and recycling facility for flume water was placed in service during 1966 in Idaho, and another installation was scheduled for use in 1967 in Oregon. Two other plants of one company are scheduled for conversion to this flume water recycle process.[25]

Present Costs of Treating Water and Wastes

Earlier in this chapter, on p. 48, it was explained that analysis of questionnaire results indicated treatment of raw water supplies is not an important cost item in the beet sugar industry. Higher and more significant costs[26] were reported in connec-

[22] Based on use of 3.5 tons of lime per day in 2,200-ton-per-day plant.

[23] Based on $15.00 per ton, incremental cost of producing additional lime in plant's existing lime kiln.

[24] Based on cost of power for pumping, $53.00 per day, 8.5 million gallons circulated per day.

[25] Hugh G. Rounds, Assistant General Superintendent, The Amalgamated Sugar Company, personal communication, April 10, 1967.

[26] The costs are 1962 replacement costs and annual operating and maintenance costs. The annual costs do not contain a capital charge.

tion with waste treatment. Forty-six plants reported equipment costs and forty-eight reported annual operating costs for waste treatment. The total investment in the equipment was $5.44 million and the total annual operating cost was $341,000. Among the plants reporting, the averages are $120,000 investment, and $7,100 annual operating cost. Averaged over all plants in the industry, the two figures are $94,000 and $5,900, respectively.

A very wide range in waste treatment costs was observed in the questionnaires. The lowest investment reported was $5,000, and four plants indicated investments of $500,000 or slightly over. Annual operating costs exclusive of capital charges also covered a wide range, the lowest among those reporting being $500 and the highest being $32,000.

As far as waste treatment is concerned, if there were absolutely no restriction on waste disposal, the long-term saving to the industry would be around $2 million annually below present levels of costs. This rough estimate is based upon certain assumptions. First, it is assumed that pulp would be dried, process water (pulp screen and pulp press) would be recirculated, condenser water would be recirculated to the flumes, and Steffens waste would be used for by-product manufacture—all for reasons of internal profitability. This means that only waste treatment and recirculation of flume waters can be considered genuine costs of waste control. It is further assumed that the investment of about $5.5 million in treatment equipment and the associated $340 thousand of operating costs would be about doubled if one takes account of the recirculation of flume water. At a 10 per cent annual capital charge (interest and amortization) the total annual cost comes out to be slightly over $2 million. In comparison with a total gross income of $500 million per year and a possible total operating cost (including the cost of the beets) of $450 million per year, this is not a striking cost item. Likewise, the investment of perhaps $11 million in waste treatment and recirculation facilities compared with a total replacement cost of all plants in the industry near $750 million indicates a reasonably small proportion associated with waste treatment. The above total waste treatment and recirculating cost converts to about 10 cents per ton of beets processed.

On the other hand, if it became necessary to eliminate nearly all of the beet sugar wastes from surface streams, a much larger investment would be required. The fairly modest investment made to the present time has resulted in the elimination of some of the most troublesome wastes, usually at moderate cost. The large volume of flume water and beet-washer water, now constituting the main waste from most plants, would require major new facilities. There does not yet seem to be enough data to predict the requirements for such a step, but the fact that some plants have invested more than a half-million dollars in waste treatment facilities, probably mainly in ponds, implies that for the nearly sixty plants in the industry, an investment of at least $25 million might be expected. Another estimate, based on about $750 thousand being required for a completely closed recirculation system in a 4,500-ton plant,[27] is approximately $30 million for the industry.[28] Extrapolation of the largest operating costs to all plants in the industry would result in an annual total waste removal expense of about $2 million.[29] This annual expense coupled with the fixed cost on the investment would make the annual total waste treatment costs approach $5 million.[30] In an industry with approximately

[27] Lloyd Jensen, personal communication.

[28] J. M. Henry, *op. cit.*, p. 231, gives detailed costs (late 1950's) of waste treatment (presumably with nearly complete elimination of pollution) in two Belgian plants. In a 1,650-ton plant, there is an investment of $120,000 (all conversions made at official exchange rates) and an annual operating expense of $17,000; a 5,300-ton plant has an investment of $350,000 and operating costs of $17,000. At a fixed capital cost of 10 per cent per annum, total yearly costs in these plants are $29,000 and $52,000 respectively. On a 2,700-daily-ton basis, the two figures would be $47,000 and $27,000. The larger of these costs, if applied to the entire U.S. beet sugar production, would be $2.7 million annual total expense.

[29] There are undoubtedly some economies of scale, but the approximate nature of these industry-wide costs does not justify attempts to refine these estimates.

[30] This is equivalent to about 25 cents total waste treatment costs per ton of beets. Phipps, in a report of British practice in 18 factories, shows costs equivalent to 14 cents per ton of beets for waste water reuse to varying degrees in these plants. Some factories practice partial reuse and some completely recirculate, so this cost is a sort of "average" of the industry. See O. H. Phipps, "National Reports, Great Britain," in *Re-Use of Water in Industry, op. cit.*, p. 236.

$35 million annual profit, this would represent a substantial expense which would result in significantly but not drastically lower profits or higher sugar prices.

Although the annual tonnage of beet sugar production in the United States has remained fairly constant for many years, it is possible that output may undergo moderate increases in the future. Reduction in raw cane sugar imports due to international political disturbances has already occurred. Appeals for removing sugar beet acreage limitations in the United States are being made by growers and processors. Actual beet growing has recently been less than programmed due in part to high grain prices and use of land for wheat production. If the interplay of these factors results in increased beet sugar production, projections of total future water demand and waste discharge might then have to be adjusted upward.

There is little doubt, however, that reductions in water and waste per ton of beets handled will more than offset any plausible increase in beet sugar production. The net effect in the industry will undoubtedly be a decrease in the total requirements for water and in the discharge of waste.

Nevertheless, as indicated above, there is some reason to expect the development of more economical waste treatment processes of several possible types. In each area or each plant, a particular total treatment process might eventually be incorporated which would permit operation at considerably lower cost than indicated above. It is not yet possible to state with any degree of certainty what these techniques and costs will be, but it can be said that the problem is amenable to solution and that the solution need not be extraordinarily expensive.

5

Economic Implications

General Conclusions

Our study has shown a clear and strong tendency for technical changes in beet sugar production to reduce waste loads and water intake per unit of product. At the same time, these changes have not so greatly altered gross water use or the quantities consumed by the industry. Mostly they have been in the direction of increased recirculation rather than the substitution of drier for wetter processes.

Reasons for the tendency toward reduced water intake and waste discharge are various. In all instances there has probably been some need felt for reducing waste discharges, combined with a growing possibility of doing so at comparatively small costs. Indeed, some processes such as pulp drying, the use of continuous diffusers accompanied by recirculation of pulp screen and press water, and the productive use of Steffens house residuals have reached a point where they are profitable, or nearly so, even in terms of the internal economics of the plant, that is, when the external-offsite costs they avoid are neglected. In any event it would require only a comparatively small external stimulus in the form of an effluent charge or an effluent standard to induce the further use of these procedures that by themselves eliminate some 70 per cent of the BOD which would be contained in the waste water from a factory using no recirculating or treatment processes.

The efficacy of these devices holds an important lesson for policy formation with regard to control of pollution from industry sources. For example, subsidies to industrial waste *treatment* in the form of rapid tax writeoff have often been suggested in the Congress and elsewhere. These could well have the effect of stimulating the use of *treatment,* in the sense of external devices for removing wastes before discharge, even though the same or better results could be achieved more efficiently by recirculation systems, internal process changes, and conversion of wastes to by-products. The same conclusion holds with respect to other major water-using and waste-discharging industries.

Another major implication of nationwide goals is that it is necessary to view the waste control problem in a dynamic rather than a static technological context. It is clear that technological change in beet sugar production has over the past few decades considerably changed the water use and waste generation characteristics of the industry. The problem is not simply to make the best use of currently available technologies, but to devise policies which foster the development of water conserving and waste reducing technology. This suggests, for example, that a control method such as an effluent charge, which puts continuous pressure on the waste discharger to reduce the amount of waste emitted, can have desirable longer-run effects on the development of technology. An effluent standard, once met, exerts no more pressure for improvement except in the sense that the costs of meeting it may be reduced over time. An effluent charge provides a continuous incentive to reduce waste discharge to any economically feasible extent, even to eliminate it. For one existing beet sugar plant, for example, there is a type of effluent charge in the form of an assessment for treating the factory effluent in the municipal system. The cost to the sugar factory is based on BOD loading, so there is an incentive to minimize discharge by recirculation and partial treatment. It is not simply a coincidence that this plant has one of the most sophisticated internal recirculation systems in the industry.

An economic factor inherent in the reduction of waste and heat discharge by recirculation is the associated reduction in cost of fresh water supply. As recirculation and internal reuse decrease the requirements for water intake, they also diminish the cost of fresh water treatment, pumping, and storage. Thus, the external benefits of pollution reduction by these methods are augmented by internal savings, so that the total value should be weighed against the cost of the recirculation system.

There now remain in the beet sugar industry, only three significant waste-bearing water streams, the external discharge of which cannot be prevented by techniques whose cost is largely offset by the recovery of valuable products or other economies internal to the plant. These are the small volume of lime cake slurry, which is extremely high in total solids and BOD; the large volume of condenser water, which carries a sizable BOD content and substantial quantities of heat; and—the most difficult and costly of all—the large volume of flume water, which is also high in total solids and BOD.

Thus, in analyzing the costs of further waste discharge reduction in the beet sugar industry, we begin with a base plant which has a continuous diffuser with associated pulp and press water recirculation system, pulp drying equipment, and which makes productive use of the Steffens residuals. From this base, we proceed to discuss the technical means and costs of mitigating the lime cake slurry, condenser water, and flume wastes to various degrees. We also trace out the implications of the various techniques for decreasing the amounts of water intake, total use, and depletion. Indeed, there are situations in which the primary motive for recirculation of condenser water, for example, is the limited availability and high cost of intake water supplies.

We feel that the summaries of cost analyses we present may provide useful information to the industry. Further, they should be valuable to water resource planners who confront the necessity of considering the costs as well as the benefits of water pollution control, and who must face the question whether it is better to provide additional water supplies for withdrawal uses and to improve the assimilative capacity of water courses to which wastes are discharged or to rely on water conservation practices in industrial plants.

Handling Lime Cake Slurry

The lime cake slurry from a typical plant of 2,700 tons daily beet capacity consists of a water stream of 250,000 gallons containing 17,500 pounds of BOD and 250,000 pounds of suspended solids. Experience in the United States and abroad indicates that, generally, the most efficient way of handling this waste stream is by means of settling ponds. None of the 58 plants in this country discharge this waste stream without some ponding, either in separate ponds or in combination with other waste water streams. Twenty-three plants have sufficient ponding capacity to hold this entire waste stream without external discharge. For our representative plant, a pond 3 feet deep and 15 acres in area would be needed if one assumes 1 foot of evaporation during the fall campaign, an equal amount of seepage, and a maximum water depth (for biological action) of 3 feet. If anaerobic conditions can be tolerated, the higher evaporation throughout the year can be utilized; with 4 feet of evaporation and an assumed 4 feet of seepage, a 6-acre pond about 10 feet deep would suffice. This evaporation rate is typical of the plains area of Colorado or Nebraska. If one assumes an evaporation during the campaign of about ⅓ foot, which characterizes Michigan, and 1 foot of seepage, the size would have to be about 3 feet by 18 acres. Construction costs, including the necessary pumps and piping for ponds of this size, would usually be in the range of $30,000 to $40,000.[1] Total costs of the pond would vary with land costs in the particular situations. At $1,000 an acre the total pond cost would be $10,000 to $15,000 higher than ponds built on land priced at say $200 per acre. Annual maintenance and operation costs can be estimated at 10 per cent.

Assuming that the pond costs fall near the top of the range, say $50,000, and that annual operating and maintenance costs are 10 per cent of the investment, and the facilities have a life of 20 years at 5 per cent interest, these costs convert to a present value of $112,500 or about $4.20 per ton of daily beet processing capacity.

Another way of viewing these costs is in terms of the cost per ton of beets processed. If it is assumed that our typical plant has an annual campaign of 100 days, it would process 270,000 tons of beets. If it is further assumed that annual operation, maintenance, and capital costs are 15 per cent of the initial investment, total annual costs would be $7,500 or a

[1] 1962 dollars.

little less than 3 cents per ton of beets processed. This is about ½ per cent of manufacturing costs exclusive of beet purchases.

Still another useful figure is the average cost of removing a ton of BOD associated with the settleable solids in the lime cake slurry. Given the same assumptions as indicated in the above paragraph, it would cost about $10 per ton removed.

As is so frequently the case with treatment processes, it is much less costly to remove even a large proportion of the waste load than to remove it all. For simple settling of solids from lime cake slurry—which all plants now practice—considerably smaller ponds are satisfactory. One-day retention, for example, which is usually adequate for nearly complete solids removal, requires about 250,000 gallons of pond capacity. This is equivalent to an acre about three-quarter foot deep. The total cost of such a small pond would seldom exceed $5,000. It has been shown in the previous chapter that this simple settling process also accomplishes the removal of most of the BOD, roughly 85 per cent. On the basis of calculations similar to those done previously for the larger pond, it is seen that the cost of the smaller pond per ton of beets processed is only about three-tenths of a cent. Similarly, it would cost about $1 to remove a ton of BOD in the settling pond, whereas the cost of removing the remaining 15 per cent of BOD by evaporation and seepage would be around $50 a ton of BOD.

Depending upon the dilution water available and the intensity of downstream use, a sufficient effect might be obtained by much shorter periods of storage or even by direct discharge of this waste stream. If it were discharged without any treatment, our representative beet sugar plant would save $750 per year or $7,500 per year, depending on whether the comparison is with 85 per cent removal or 100 per cent removal of BOD. The relationship between removal capacity and cost is described by an exponential function, so that it is possible to estimate the cost of any intermediate size of pond. The longer the retention period, the greater the degradation of BOD, but at a much slower rate.

It is evident from the figures on ponding in the industry (cited above) that a severely polluting waste stream has been greatly diminished in all plants and virtually eliminated in some. It is clear also that total elimination of this waste stream is possible but would involve substantial cost. Nationally, this waste stream before any treatment accounts for 20 to 30 per cent of the BOD generated in the various beet sugar plant processes which have previously been described. (See Table 3.) The final effluent now accounts for 8 to 10 per cent of the BOD which is still being discharged from this industry to the nation's streams.

Treating Condenser Water

Our representative plant will produce a condenser water stream of about 5.4 million gallons per day. This stream will contain about 2 to 2.5 billion Btu of heat and 1,800 pounds of BOD. The heat discharge results in a temperature increase in the condenser water of 50 to 60 degrees. Costs of eliminating each of these contaminants may be separately considered.

The cost of recooling condenser water may be estimated from data on design wet-bulb temperatures in a particular locality and from published information on cooling tower costs.[2] Using typical conditions, the above cooling requirements could be met by a cooling tower investment of about $100,000. Operating and capital costs would total about 2.5 to 3 cents per thousand gallons circulated, so this practice would become economical (in the absence of other in-plant costs or economies accruing) if the total cost of fresh water delivered to the plant cooling system is as much as 3 cents per thousand gallons.[3] Similarly, if an effluent charge of say 5 cents per million Btu discharged were to be assessed, recirculation through cooling towers at a cost of 6 to 7 cents per million Btu (equivalent to

[2] Brian Berg, Russell W. Lane, and Thurston E. Larson, "Water Use and Related Costs with Cooling Towers," *Journal of the American Water Works Association*, Vol. 56 (March 6, 1964), p. 311. Also, Paul H. Cootner and George O. G. Löf, *Water Demand for Steam Electric Generation* (Washington: Resources for the Future, Inc., 1965).

[3] In this total would be interest charges and amortization of investment in wells, intake facilities, pumps, raw water treatment equipment, as well as the operating cost of these systems.

2.75 cents per thousand gallons under the chosen conditions) might well be the chosen alternative. Or if an effluent standard were to be imposed which permitted discharge of only a portion of the heat produced in the plant, recirculation at costs of this magnitude would permit compliance. Thus, various heat removal levels can be accomplished—from the total removal of 2 billion Btu per day for about $120 to $140 per operating day ($12,000 to $14,000 per year) down toward fairly low cooling rates at approximately proportional costs. These total heat removal costs are roughly equivalent to 5 cents per ton of beets handled. The technology and economics of cooling towers turns out to be such that if a stream is recirculated at all, it will be recirculated in a more or less closed cycle, which substantially eliminates any heat discharge in the waste water from the plant.

Depending upon the costs of land, it may be less costly to recirculate condenser water after cooling in ponds. For our "representative" plant, a pond of about 20 acres size would be necessary to effect cooling of condenser water containing 2 billion Btu per day, assuming the plant is located in a relatively dry and cool climate such as in the mountain states. Applying the same assumptions concerning land costs and construction expenses indicated in the discussion of lime slurry ponds, a typical cost for such a pond would be about $50,000. In contrast with cooling towers, the ponds have small maintenance and operating costs. Pumping and capital costs can be approximated as 1.3 cents per thousand gallons circulated, equivalent to 2.6 cents per ton of beets, for a total cost of $7,000 per year.[4] It is seen that, under the conditions assumed, the use of a pond is considerably less costly than the use of a cooling tower. However, if atmospheric conditions are less favorable (as in central and southern California) or if land costs are much higher, the cooling tower may be the more economical alternative. Virtually complete recirculation of the condenser water, except for a minor fraction necessarily discharged

[4] Computed by assumption of 10 per cent annual cost on pond investment, 50 feet equivalent pumping head, power at 1 cent per kwh, 2,700-ton beet plant, 100-day campaign.

to avoid buildup of impurities, would reduce the water intake of our representative plant by about 40 per cent below what it would be if none of the water streams (except pulp screen and press water) were recirculated, and it would reduce heat discharge substantially to zero. The cost of water supply, although usually moderate (Chapter 4, pp. 48–49), would be decreased somewhat less than proportionally.

Another cooling system involves use of a spray pond, which might be considered a compromise between the large area of a pond and concentrated cooling in a tower. By use of an array of nozzles in a basin of an acre or so, warm water may be effectively cooled at moderate cost, if atmospheric variables —particularly with respect to dependable wind velocities—are suitable. Costs are usually less than with cooling towers but are higher than ordinary ponds, unless the latter have to be unusually large or are on costly land.

Using the costs cited above, it appears that the cooling water streams of all beet sugar plants in the United States could be recirculated by means of towers for a capital investment of about $6 million and annual operating costs of $360,000. Capital plus operating costs of $1 million per year would be incurred. Since nearly all of the Pacific Coast plants already employ cooling towers, additional investment by other plants for comparable facilities would represent about $4.5 million.

If ponds for cooling and stabilization were used in all plants not now employing cooling towers, about $3.2 million investment would be required. Total recirculation of the condenser water would have the effect of reducing the intake water of beet sugar plants across the country by 35 to 40 per cent and it would reduce the heat load going to streams by about 100 billion Btu per day. If daily heat discharges of this magnitude, namely 2 billion Btu per typical plant, were only partially removed in recirculation systems, costs would be nearly proportionally reduced, whether by cooling towers or by ponding. Thus, a near linear relationship between heat removal and cost prevails. The function relating these variables may be represented as:

$$C = K \times H,$$

where C is the annual cost of heat removed in dollars, K is about $7,000 for ponding or $13,000 for cooling towers (adjusted for local atmospheric conditions as required), and H is the fraction of the heat removed from the condenser water. Analysis performed for the publication, *Resources in America's Future,*[5] serves as the basis for inferring that recirculation or lack thereof has no substantial effect on the overall amount of water evaporated or "lost" in the condenser water function, when considered in terms both of water within plant and after discharge to streams.

The second problem associated with condenser water discharge is its BOD content, resulting from sugar carry-over, approximately 1,800 pounds per day from our representative plant. Neither cooling towers, nor spray ponds effect a significant reduction in BOD,[6] so recirculation through one of these units results in a steady increase in sugar concentration, unless some new water is continually added and a portion of the circulating stream diverted. Fermentation of sugar in this stream may occur, so chlorination may also be required.[7] If recirculation is sufficiently intensive to permit no more than 5 to 10 per cent diversion, that is 100 to 200 gallons per ton of beets, this stream can usually be totally used as makeup water to the diffuser system. Thus, there need be no condenser water discharge, and both heat and BOD are entirely removed. The cost of this further recirculation system, if chlorination cost is arbitrarily assigned to BOD control rather than heat control, may be estimated as one-half cent per ton of beets handled, equivalent to about one cent per pound of BOD removed or $1,200 per year. This estimate is based on the assumption that equipment for chlorination and pumps and piping for supplying water to the diffuser circuit would cost approxi-

[5] Hans H. Landsberg, Leonard L. Fischman, and Joseph L. Fisher, *Resources in America's Future* (Baltimore: The Johns Hopkins Press for Resources for the Future, Inc., 1963).

[6] However, see possible development mentioned on pp. 69–70.

[7] Chlorine addition at the rate ot 3 ppm., equivalent to 135 pounds per day in a 2,700-ton plant, is considered adequate for controlling fermentation in condenser water. See O. H. Phipps, "National Reports, Great Britain," in *Re-Use of Water in Industry* (London: Butterworth & Co., 1963), p. 236.

mately $10,000. Thus, if a cooling tower is used, complete condenser water treatment for heat and BOD removal would involve about $110,000 investment and $15,000 total annual cost. Under favorable conditions, where relatively cheap cooling ponds can be used, investment and annual costs should be about half these figures.

If the blow-down (portion of condenser water discarded) is not reused, its BOD may be reduced or eliminated by ponding, thereby oxidizing BOD and decreasing water content by evaporation and seepage. If 30 to 60 days are required for nearly complete stabilization of this effluent, 10 per cent of the condenser water would require a pond volume of 16 to 32 million gallons, or 50 to 100 acre feet. A combination of anaerobic and aerobic ponding, with an average depth of 3 feet would thus require pond areas of 17 to 35 acres. In most instances, this acreage should be adequate for cooling purposes also, except in the California plants. The previously stated costs of heat removal by ponding should then also cover the expense of substantially complete removal of BOD from condenser water if the ponding method is feasible at the location concerned.

So far it has been assumed that the condenser-water stream, if it is recirculated at all, would be recirculated to once again perform the cooling function. Actually, it is much more common practice for part or all of the condenser water to be reused for fluming and beet washing. The flume waters are then partially or wholly discharged, often through settling ponds. Thus, reuse of condenser water in the flumes has the effect of reducing intake in the same way as does recirculation of condenser water through cooling towers or ponds. Also, reuse of condenser water in the open flume system has an effect of reducing heat load going to the streams through the cooling of water by evaporation and the thawing of frozen beets. However, the final discharge temperature of this water is usually considerably above normal stream temperature, so thermal pollution is still a consideration. Reuse of condenser water for fluming does not reduce the discharge of BOD in this waste water, because the sugar picked up in the condensers carries directly through the subsequent operations. However, the reduction of total BOD discharged from these two operations by treatment of one stream of higher BOD concentration may be more effectively accomplished than by separate handling of the dilute condenser water. It should be apparent, however, that unless a completely closed system is being used for these two streams, the option of reusing condenser water in the flumes is not open to a plant that practices high recirculation of flume water, because excess water would accumulate in the fluming system.

Treating Flume and Beet Washing Water

The evidence we have obtained from the industry indicates that handling of the flume and beet washing waters is the most difficult and expensive remaining problem for the industry. It appears that any plant operator wishing to reduce waste discharge issuing from this source, would find it economical to engage in extensive recirculation of the flume water stream. One option would be completely closed circulation to eliminate any discharge, but even if some of the stream were discharged from the circuit and treated before joint discharge, extensive recirculation would reduce treatment costs by decreasing the water volume finally treated. Savings would also result from reduced intake. Based on the current discharge of all wastes by the beet sugar industry (Table 6), complete removal of BOD from flume waste would reduce BOD discharge by about 55 per cent. On the basis of practices assumed for our "representative" plant, elimination of this one waste from effluents containing all the BOD in condenser water and lime slurry overflow (the latter from effective settling ponds) would effect a 75 per cent reduction in BOD. Suspended solids reduction, on these two bases, would be nearly total, because this waste is currently the only significant source of suspended solids pollution. Indeed, if our representative plant succeeded in totally impounding its lime slurry, recirculating its condenser water by one of the alternatives previously described, and shutting off the waste stream resulting from fluming, it would have achieved a complete absence of waste discharge to water courses. There are several plants in this country which have virtually attained this result by variations of the two general methods, namely, extensive recirculation and waste treatment.

One method—used in one plant—is complete closed recirculation of the flume water stream through a clarification process, in combination with condenser water reuse for fluming, cooling in a spray pond, and recycle through condensers in an integrated water system. Another essentially closed loop used in this plant is diffuser supply from pulp screen and press waters and makeup from a small fresh water supply. The major factor inhibiting the use of this technique, in addition to its cost, is the accumulation of various impurities in the recirculating stream. This makes it necessary to provide for removal of suspended solids by clarification, either in a settling basin, or in a mechanical clarifier. Pretreatment of this stream, by addition of lime or other chemical, aids coagulation and settling of the very fine particles of soil and other suspended matter. Also, to control odors and fouling of equipment by decomposition of sugars and other dissolved components, chlorination[8] and other chemical treatment of the recirculating stream may be necessary.

Because of the need for some net additions of fresh water to the recirculation system, there may be a gradual accumulation of liquid wastes that must be disposed of in some manner. The one plant using this system discharges its wastes to the municipal sewage treatment plant, predominately after the campaign. Experience at this plant indicates that it would cost about $500,000 for equipment and $30,000 annual operating costs for our representative plant to institute such a system. This would result in a total annual cost of $80,000 to $100,000. The cost of partial recirculation, which might be achieved by bleeding the stream and supplying makeup water from another source, would be proportionately lower. These costs are based on an assumption that a mechanical clarifier would be a

[8] O. H. Phipps, *op. cit.*, p. 236, indicates typical dosage of 10 ppm. chlorine, equivalent to about 0.2 pound per ton of beets or 540 pounds per day (about $20 cost) in a 2,700-ton plant.

component of the recirculation facilities. Such a system requires less land area than one employing settling ponds, but some pond capacity is needed to provide sufficient water storage volume for fluctuations in requirements. However, in those instances where land costs are comparatively low, a saving might be made by substituting a settling pond for the clarifier. For our representative plant, a pond of 30 acres capacity and $60,000 land and construction cost (including the necessary piping and pumps) would probably be necessary. If about 25 acres of pond are required for settling, and if land costs are higher than about $2,000 per acre, it might be more economical to construct the mechanical clarifier than a $75,000 to $100,000 pond.

As described in a previous section, a new ponding system characterized by virtually complete recirculation, short retention times, and lime addition may prove widely applicable. If further trial leads to such a conclusion, complete treatment of flume waters would be considerably cheaper than in the plant described in the foregoing paragraph. A capital cost of about $200,000 and annual operating expenses of $13,000 would place the total annual costs of completely eliminating flume and washer wastes at about $35,000. Until additional development and testing are undertaken, however, this process and its apparent costs should not be considered established.

Another method which might be used to ameliorate the polluting effects of this effluent stream, perhaps reducing it to a point where it produced negligible external damages, would be to provide storage for the stream and discharge it during periods of high river flows. Maximum flexibility would be achieved by constructing a pond of sufficient size to store the stream during the entire campaign. For our representative plant (employing no recirculation or reuse of condenser water or flume water) a large pond of 4,000 acre-feet capacity and $0.5 to $1 million typical construction cost would be necessary. This very high cost would ordinarily preclude this alternative, unless perhaps a natural pond could be utilized, or the terrain made it possible to construct a much cheaper reservoir. Operational cost of such a facility would be a minor factor, so the total annual costs of complete ponding of wastes from a plant using no recirculation of condenser water or flume water, with regulated discharge during periods of high stream flow, could range up to $100,000 or more, depending on numerous factors. It is evident that partial or even complete recirculation, in combination with a smaller ponding system, would usually be more economical.

A system of partial recirculation of flume water through ponds (with or without mechanical clarifiers) and partial recirculation of condenser water through cooling ponds or cooling towers, the balance going to flumes, may be coupled with regulated discharge. By such means, necessary pond volumes can be reduced by factors of ten or even more. Thus, 400 acre-feet of storage would usually be sufficient for storing all the liquid wastes during a 100-day campaign of a 2,700-ton plant if fresh water makeup to the condenser water system is 10 per cent of use and if fresh water makeup to fluming and washing is also 10 per cent. The characteristics of this system and its typical cost are not greatly different from the total recirculation system outlined above. Possibly less stringent measures for controlling in-plant effects of BOD in recirculating streams, but more expensive ponding facilities, would be involved. Representative annual costs would probably lie in the $50,000 to $100,000 range.

Another method for reducing, although not completely eliminating, the waste discharge from this stream, is by means of treatment with a trickling filter. In general, this has seemed to be less economical than the alternative methods and would usually involve extensive recirculation in order that the filter can be of reasonable size. It appears that considerable pond capacity will also have to be provided in plants employing this system, mainly to reduce fluctuations in flow and composition of wastes delivered to the trickling filter and to lengthen the treatment period.

If the flume water were recirculated without appreciable discharge, the water intake of our representative plant would be reduced by about 7 million gallons per day. If both the condenser water and the flume water streams were recirculated, the daily water intake of our plant would be reduced from about 13 million gallons to an almost negligible 700,000 gallons of makeup water.

Before proceeding to a more systematic estimation of the functional relationship between costs and the reduction of water intake and waste flows, it should be noted that some other methods have found application in various parts of the world. For example, the spraying of waste waters on meadows appears to be a comparatively common practice in Germany. In certain quite arid areas and where the soil is porous, it is feasible to store in ponds the waste waters that remain after considerable recirculation. There they evaporate and infiltrate into the ground, so that there is no surface discharge. Several mills in this country use the latter method. There may also be opportunities to substitute dry processes for those which are presently wet. For example, the lime cake might be removed in a semidry state rather than converted into a slurry, or the beets might be transported to the slicers by mechanical means rather than by a stream of water. All of these procedures may find application in particular instances. Since they do require either some improvements in technology or somewhat specific circumstances, we have chosen not to include them in our summary statements of waste reduction possibilities and costs.

Cost of BOD Reduction

We will now proceed to the matter of estimating a cost function for our representative plant. Our aim is to derive a reasonable function which relates cost to BOD reduction. At the same time we will indicate the effect on total water intake. It will by now be clear to the reader that there are a great many possible combinations of techniques for reducing waste discharge, and therefore the techniques that we specify in estimating our cost function must be considered no more than illustrative of those that a representative plant might use. We begin once more with a base plant that has already installed a continuous diffuser to which it recirculates pulp screen and press water, a plant that has pulp drying equipment, and that makes productive use of its Steffens house residual. We then indicate how the plant might achieve various levels of BOD reduction by handling its lime cake slurry, its flume waters, and finally its condenser water—the latter also involving the elimination of waste heat discharge. The first step we specify for the reduction of BOD discharge is the settling of lime cake slurry in a pond with one-day retention time, a treatment now equalled or exceeded by the entire industry. A second step is the ponding of both the lime cake slurry and the flume waters with one-day retention time. The third step is the complete withholding from discharge of the lime cake slurry, including the clarified fluid portion, while at the same time ponding the flume waters with a one-day retention period.[9] The fourth step is complete withholding of both of these waste streams, with the flume waters being recirculated in a closed system. The final step is virtually total recirculation of flume and condenser water, in combined or separate systems, with total ponding of lime wastes and excess water from fluming and condensing, and with cooling of condenser water in cooling towers; the small net waste outflow is handled by municipal plant or by regulated harmless discharge. Since there are obviously a great many possible combinations of these techniques, we assume it is legitimate to interpolate between the points established for the specific steps, thereby obtaining a continuous function relating costs to BOD reduction. (Figure 6.) From this cost function we derive an incremental cost function which is indicated in Figure 7. All costs are indicated in terms of present values.

The present values were computed by adding capital investment and the total operating costs over an assumed twenty-year life at 5 per cent interest. The figure thus represents the total funds required to be set aside at the present time for building the facilities and operating them for twenty years.

The data for incremental cost of BOD removal, shown in

[9] In some instances, the additional treatment measures indicated in steps 2 and 3 may be economically reversed, especially in areas where seepage and evaporation rates are high, viz., second step is complete retention of lime cake slurry and third step is the addition of one-day settling of flume wastes.

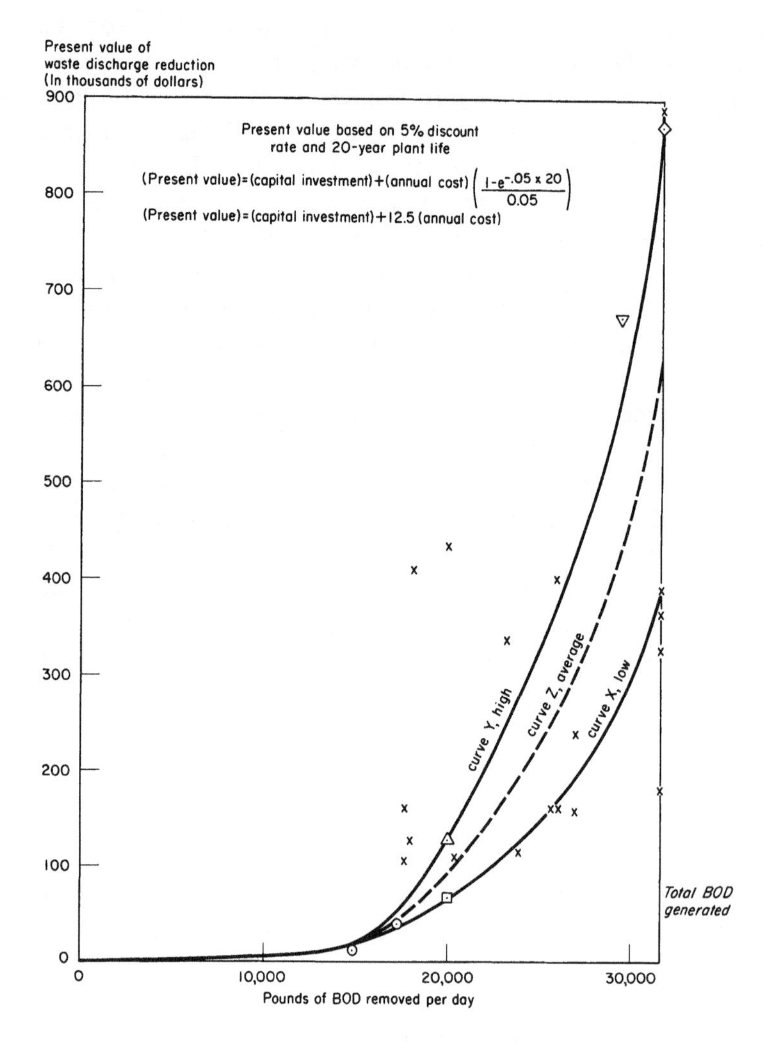

Figure 6. Cost of reducing BOD content of lime, flume, and condenser water wastes from 2,700-ton-a-day beet sugar plant. (The data for this figure are of two types. The points marked by *X* are actual calculations based on costs reported in questionnaires and estimates of BOD removal, converted to the 2,700-ton basis by simple proportion. The points shown by circles, square, triangles, and the diamond shape are calculations based on general estimating procedures outlined in the chapter, by use of the several steps of waste treatment in the hypothetical plant. Curves *X* and *Y* are drawn through the "low" data and "high" data respectively. The dashed curve *Z* is an average.)

Figure 7, have been obtained from Figure 6 by plotting the slopes of the present value curves in Figure 6 against the BOD removals. That is, the increase in cost of removing a pound of

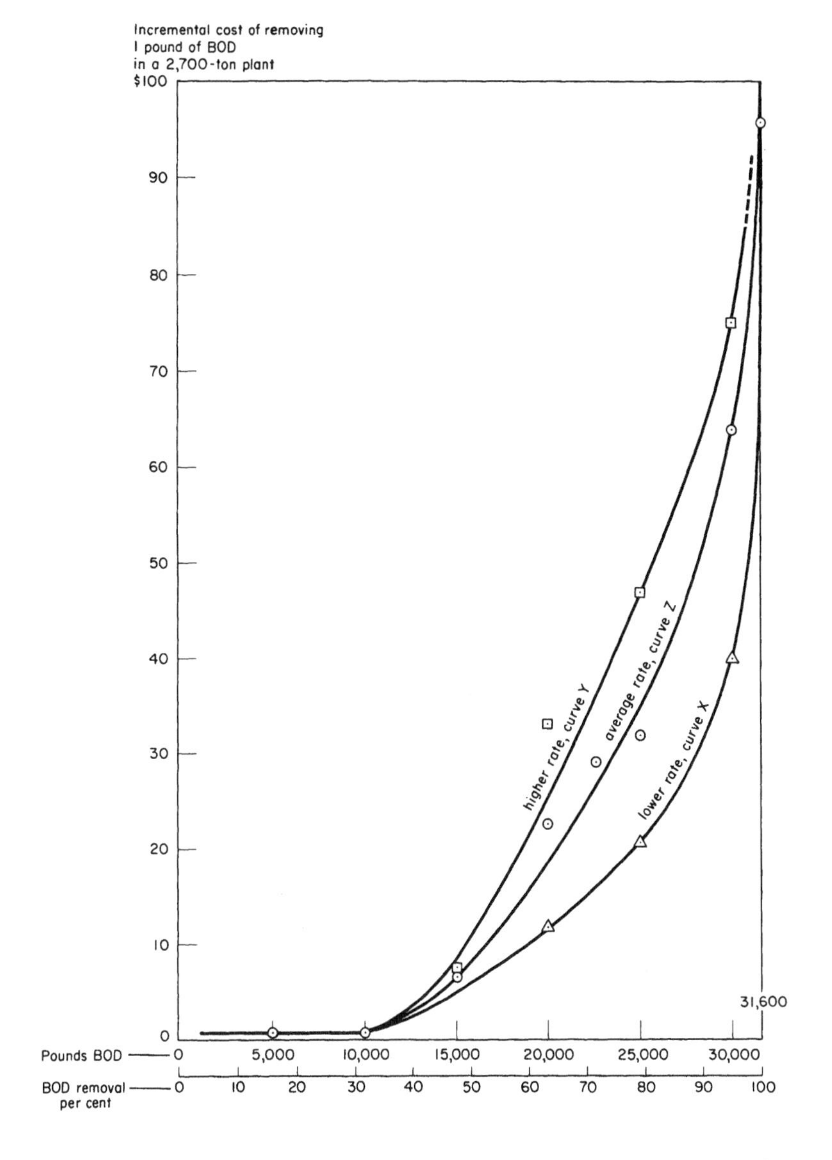

Figure 7. Incremental cost of reducing BOD content of lime, flume, and condenser water wastes from 2,700-ton-a-day beet sugar plant.

BOD was obtained by dividing the increase in present value by the change in quantity removed.

In deciding upon a level of waste control which would minimize the overall costs associated with the disposal of wastes, including external costs, an incremental damage cost function would have to be derived. A hypothetical function of this kind is shown in the next illustration, Figure 8, and also the point of optimal waste discharge is indicated—optimal in this instance meaning the cost minimizing point, that is, the point where the sum of internal wastes control and external damage costs are at a minimum. In an actual effort to optimize waste discharges in an entire river basin system, the calculations would become much more complex, involving a simultaneous consideration of numerous points of waste discharge in water use, and possibly opportunities for streamflow regulation or other measures to improve the assimilative capacities of streams. It is quite possible that some analytical procedure, such as system simulation, would be necessary to establish optimum amounts of waste discharge for each point in the basin. Nevertheless, even such a sophisticated approach would find waste reduction and damage cost functions indispensable, and the criterion for optimality indicated in the simple illustration above would continue to hold.

Concluding Comments

We hope that the information and analysis of water utilization in the beet sugar industry will prove of interest and utility to the industry and to those charged with the development and management of water resources. The latter will find, we believe, that the study has shown the folly of making projections of water "requirements" and waste loads by applying coefficients to projected industrial production (an all too common procedure). Those who neglect economics in the analysis and projection of industrial water utilization do so by risking grievous error.

Also, we believe that the study clearly illustrates that incentive programs aimed at stimulating treatment in "end of

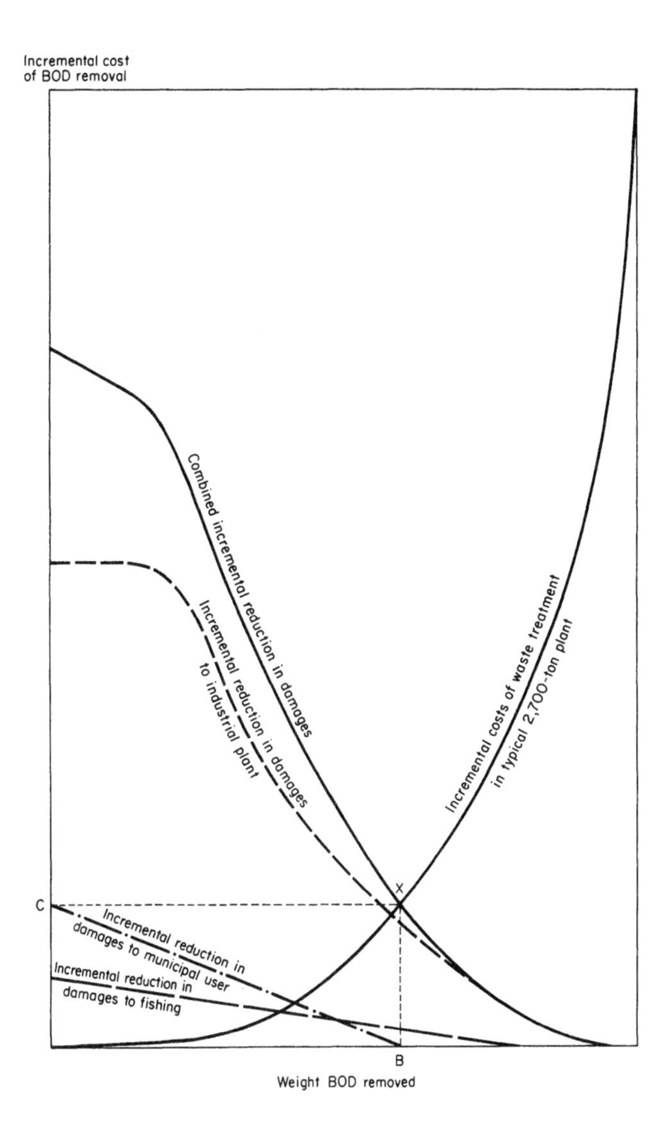

Figure 8. Reduction in BOD discharged in lime slurry, flume water, and condenser water from 2,700-ton-a-day beet sugar plant. (Example shows that optimum waste treatment—for minimum internal and external cost—corresponds to point *X*, where *B* pounds BOD are being removed, at cost, for the last increment of BOD removed, of *C* dollars per pound BOD. At this point, if the actual values for this graph are known, Figure 6 can then be used to obtain the total cost of this level of treatment on a "present value" basis.)

the process" plants may lead to grossly inefficient measures for waste reduction. While this study was being written, there was much sentiment for federal programs to subsidize industrial waste "treatment" and many bills to this effect were introduced into the Congress. This approach invites serious rethinking in view of the many economical opportunities for waste discharge reduction via in-process changes and recirculation. Indeed, in the beet sugar industry the reduction of waste discharge can best be accomplished by higher rates of recirculation and associated water quality control. A zero waste discharge plant would optimally involve a completely closed recirculation system.

The information in this study should permit planners and those responsible for public policy to make improved estimates of water utilization and necessary economic costs for various levels of waste reduction now and in the future.

Appendix A

Water Usage Relationships by Regression Analysis

by

Robert M. Steinberg and Betty Duenckel
Resources for the Future, Inc.

The water-using characteristics of plants in the beet sugar industry were related by multiple regression analysis to plant size, age, and the presence or absence of certain types of process characteristics. In some cases a water-using characteristic was also important as an independent variable. For example, gross water use was strongly related to recirculation and vice versa. In general, statistically significant relationships were discovered, the signs of the regression coefficients were consistent with intuitive judgment of cause and effect, and there was no trouble with multicollinearity.[1]

Regressions were run for experimental purposes on various subsets of the total fifty-eight plants. The number of plants in regression equations reported below may be:

58–Total plants

53–Excluding plants with no information on flume use and process use

48–Excluding plants in California

46–Excluding plants in California, and with no information on flume use and process use

30–Excluding plants in California, excluding plants greater than 2,554 tons capacity

18–Excluding plants in California, excluding plants less than 2,554 tons capacity.

The variables in the analysis are defined below and the mean and standard deviation for the fifty-eight plants in the analysis are shown.

[1] Multicollinearity describes a condition of the data where independent measures of the contribution of each variable are difficult to obtain because of high correlations within the set of independent variables.

		Mean	*Standard deviation*
X_1	Plant size in thousands of tons of beets per day	2.724	1.308
X_2	1 for plants in California, 0 for other plants	0.17	0.38
X_3	Gross water use, thousands of gallons per ton of beets	4.7	1.5
X_4	Recirculated water use, thousands of gallons per ton of beets	2.2	1.3
X_5	Recirculation as per cent of gross use, tens of per cent	4.51 (45.1%)	1.99
X_6	Gross non-flume water use,[2] thousands of gallons per ton of beets	2.7	1.1[3]
X_7	Age of plant in 1962, tens of years	4.60 (46.0 years)	1.57
X_8	1 for use of Steffens Process, 0 for non-Steffens plants	0.38	0.49
X_9	1 for use of continuous diffusion, 0 for cell-type diffusion	0.72	0.45
X_{10}	1 for use of lagooning, 0 for no lagooning of wastes	0.24	0.43
X_{11}	Gross flume water use, thousands of gallons per ton of beets	2.1	0.8[3]
X_{12}	Intake, thousands of gallons per ton of beets	2.5	1.0

Identities

$$X_3 = X_6 + X_{11}$$
$$X_{12} = X_3 - X_4$$
$$X_5 = 10X_4/X_3$$

[2] This variable includes all categories of water use other than flume and beet washing use (i.e. diffusion, cooling, etc.).

[3] Mean and standard deviation are based on 53 plants for which data are available.

Some general conclusions follow from the analysis of all the regression results. Results do not differ very significantly from one subset to another subset of plants with one important exception: Analysis of the residuals[4] of the regression fits with forty-six to fifty-eight plants showed similar phenomena regardless of the dependent variable. The plants of smaller capacity had widely oscillating residuals, the plants of larger capacity had residuals oscillating within a much smaller range. This pattern of residuals suggested an inherent unpredictability in the water-using characteristics of the smaller plants. This was confirmed by dividing the plants outside California into two groups above and below the mean plant size. For some of the dependent variables there was no significant regression relationship for the group of smaller plants. On the other hand, for the group of larger plants the square of the multiple correlation coefficient (percentage of total variance explained) was considerably higher. A table of differences in R^2 is shown below.

Dependent variable	R^2 *for total (48 plants)*	R^2 *for large plants (18 plants)*
X_3 –Gross water use	.4286	.8461
X_{12}–Intake	.3480	.4547
X_4 –Recirculation	.7219	.9720

The signs and magnitudes of the regression coefficients were similar whether for all plants or large plants only. Therefore the regression equations reported below are for the total number of plants with the warning that small plants fit the relationship less well, and large plants better than the average.

One other conclusion comes from the analysis of subsets of the original data set. The set of independent variables certainly does not describe all factors affecting the water-use characteristics of the plants. It was found that California plants equal in respect to other explanatory variables with plants outside California tended to have gross use of 1.6 thousand gallons per ton less, with less intake of 0.9 thousand gallons, and greater recirculation of 0.7 thousand gallons per ton. This probably reflects the difference in

[4] Residuals are calculated for each plant as the difference between the actual dependent variable and the predicted value of the dependent variable derived from the right-hand side of the equation.

water availability between California and the rest of the nation (not measured by the set of independent variables).

The regression equations for all plants are shown below. The F value[5] for the particular independent variable is shown in parentheses below the value of the regression coefficient. For the number of plants in the analysis the degree of confidence and the F values are shown below:

Confidence	*F value*
99%	7.17
95%	4.03
90%	2.81
80%	1.68
70%	1.10

Gross water use

$$X_3 = 1.73 - \underset{(15.8)}{1.57X_2} + \underset{(26.5)}{.412X_5} + \underset{(14.0)}{.358X_7} - \underset{(9.7)}{1.10X_{10}}\ ^{6}$$

$$R^2 = .4865 \qquad N = 58$$

Intake

$$X_{12} = 2.32 - \underset{(2.9)}{.149X_1} - \underset{(9.4)}{.864X_2} + \underset{(8.1)}{.201X_7} - \underset{(11.4)}{.806X_{10}}$$

$$R^2 = .4869 \qquad N = 58$$

Recirculation use[7]

$$X_4 = -1.52 + \underset{(4.9)}{.742X_2} + \underset{(100.7)}{.827X_3} - \underset{(6.9)}{.207X_7} + \underset{(1.7)}{.353X_8} + \underset{(2.7)}{.463X_9} + \underset{(10.3)}{.867X_{10}}$$

$$R^2 = .7162 \qquad N = 53$$

[5] The higher the F value the less probable is the possibility that the linear relationship between independent and dependent variable was mere happenstance of the statistics used. Even if no linear relationship existed (the null hypothesis), a non-zero coefficient would be extracted from the regression analysis. These regression coefficients for the null hypothesis have a known probability distribution given all other variables in the analysis, the number of observations, and a few assumptions. The larger the absolute value of the coefficient the less probable is the null hypothesis; 99 per cent confidence indicates a 1 per cent chance that no linear relationship exists, 95 per cent confidence indicates a 5 per cent chance, etc.

[6] The regression coefficients of this equation may be interpreted in a literary way as follows. All numerical values are scaled to thousands of gallons per ton of beets.

1.73–Amount of gross water use not otherwise explained when X_3, X_2, X_5, X_7, and X_{10} are at their mean values

Recirculation as per cent (X_5 is scaled here to whole number of per cent)

$$X_5 = 30.7 + \underset{(4.1)}{14.3X_2} + \underset{(16.1)}{8.60X_6} - \underset{(5.9)}{4.01X_7} + \underset{(1.9)}{7.70X_8} + \underset{(0.7)}{5.06X_9} + \underset{(8.6)}{16.8X_{10}}$$

$R^2 = .4228 \qquad N = 53$

Although no statistically significant relationship was discovered relating gross water use to plant size or the use of the Steffens process, somewhat surprisingly, gross flume water use was related to these two variables.

Gross flume water use

$$X_{11} = 2.69 - \underset{(10.9)}{.270X_1} + \underset{(4.5)}{.492X_8}$$

$R^2 = .1952 \qquad N = 53$

Finding such a correlation between the Steffens process and gross flume use highlights the limited interpretation that should be made of these regression results. The equations do not necessarily express cause and effect. One does not expect to increase gross flume use by .492 thousand gallons per ton by using a Steffens process. Rather, the higher gross flume use tends to be associated with those plants that use the Steffens process.

1.57–Amount of gross water use by which California plants fall below rest of U.S. plants given equal recirculation percentage, age, and use of lagooning

.412–An increment in gross water use associated with an increment of 10 per cent in recirculation

.358–An increment in gross water use associated with an increment of 10 years in the age of the plant

1.10–Amount of gross water use by which plants that use lagooning fall below plants that do not, given equal recirculation percentage, age, and location relative to the state of California.

[7] No significant statistical relationship was discovered between recirculation use, X_4, and plant size, X_1, for the U.S. A subset equation excluding plants in California found a statistically significant negative relationship between recirculation use and plant size. The explanation of these two findings may be that outside California, large plants were located in general only where abundant supplies of water were available obviating the need for recirculation, whereas in California an essential ingredient for large plants was a high degree of recirculation, thus masking the relationship discovered in the subset equation.

Appendix B

QUESTIONNAIRE

Industrial Plant Water Supply and Waste Disposal
for the
National Technical Task Committee on Industrial Wastes

* * * *

Modified for Use in Study by Resources for the Future

SECTION A **GENERAL**

1. Company name ______________________________

2. Location of plant ______________________ | ______________________
City State

3. In what year was plant built? ______________________

4. Typical dates of campaign, from ______________ to ______________

5. Typical annual beet tonnage processed ______________________

Beet tonnage processed in 1962 ______________________

Typical 24-hour beet tonnage processed ______________________

6. If plant includes Steffens process, typical annual tonnage

molasses processed ______________________

Molasses tonnage processed in 1962 ______________________

Typical 24-hour molasses tonnage processed ______________________

SECTION B WATER USE

1. Gross Fresh Water Intake (Approx. 1000 gal/day)

1962	1949 (or first year of production)

2. Source of Fresh Water (Approx. 1000 gal/day)

Wells		River or Lake	
Municipally owned		Municipal supply	
Private water company		Private water company	
Plant owned		Plant owned intake	

3. Brackish or Salt Water (1000 gal/day)

4. Additional process and cooling water (1000 gal/day) which would have been required if no water was recirculated or reused:________________

SECTION C WATER AND WASTE TREATMENT

1. Is water treated in plant facilities before use? ☐ No ☐ Yes (If Yes, what per cent of total is treated?__________________% and what types of treatment are provided?)
   ☐ Complete demineralization ☐ Filtration ☐ Settling or sedimentation
   ☐ Coagulation ☐ Softening ☐ Other (Specify)
   Estimated replacement cost of facilities to treat water before use:

   $____________________________________

   Estimated annual expenditure (operating cost exclusive of capital charges) for water treatment: $__________________________

2. Are final waste waters treated before discharge? ☐ No ☐ Yes (If Yes, what per cent of total is treated? __________________% and which of the following processes are used?) (Please underscore)

Screening	Chemical oxidation or reduction
Sedimentation	Elutriation
Centrifuging	Emulsion breaking
Filtration	Distillation and stripping
Flotation (air)	Evaporation
Flotation (gravity)	Incineration
Flocculation	Chemical coagulation
Neutralization	Dewatering
Biological filtration	Fermentation
Activated sludge	Disinfection
Anaerobic digestion	Stabilization ponds
Extraction	Regulated discharge
Wet oxidation	Other (specify)

Estimated annual expenditure for waste treatment in item 2 above: $____________

Estimated replacement cost of facilities to treat waste water: $________________

SECTION D-I

DISCHARGE, 1962

1. Approximate total discharge of waste water (1000 gal/day)

	Flume and beet washings		Pulp waters		Lime cake slurry		Cooling and condensate		Steffens	
2. Waste water disposition to:	treated	untreated	treated	untreated	treated	untreated	treated	untreated	treated	untreated
River or stream										
Fresh water lake										
Municipal sewer										
Ocean or salt water body										
Earthen basin										
Spray irrigation										
Underground disposal										
Other (specify)										

SECTION D-2 **RECIRCULATION AND RE-USE** (quantities in 1000 gal/day), 1962

Source of water			Quantities re-used per day for:				
	Condition	Total quantity	Flumes and beet washing	Diffusers	Filter washing	Cooling water	Miscellaneous
Flume waters and beet washer water	treated						
	untreated						
Pulp screen and press water and silo drainage	treated						
	untreated						
Lime cake slurry	treated						
	untreated						
Condensate	treated						
	untreated						
Steffens waste	treated						
	untreated						
Cooling water	treated						
	untreated						

SECTION E WATER AND WASTE QUALITY, 1962 OR 1963

		BOD		Total solids		Suspended solids		pH	Temp. °F
		ppm	lbs/day	ppm	lbs/day	ppm	lbs/day		
Raw water supply									
Flume waters and beet washer water	before treatment								
	after treatment								
Pulp screen and press water, silo drainage (if any)	before treatment								
	after treatment								
Lime cake slurry	before treatment								
	after treatment								
Cooling water	before treatment								
	after treatment								
Steffens wastes (if any)	before treatment								
	after treatment								

SECTION F WATER PLANS

1. Indicate which of the following is contemplated:

a. Water Supply	b. Waste Water Disposal
☐ Add new source or enlarge existing source. Estimated cost $________	☐ Construction of new or additional waste treatment facilities. Estimated cost $________
☐ Construction of new or additional facilities for treating raw water. Estimated cost $________	☐ Relocation of outfall lines. Estimated cost $________
☐ None	☐ Changes in process which will reduce amount of industrial waste. Estimated cost $________
	☐ None

SECTION G WATER RATIO

1. If possible, calculate amount of water this plant uses in producing a unit (ton, bbl, case, etc.) of finished product:

Product	Unit of Product	Gallons per Unit

SECTION H COMMENTS

SECTION I DISPOSAL OF BEET PULP AND STEFFENS WASTE IN 1962

A. Pulp

Quantity sold as wet pulp, tons per year__________

Quantity sold as ensilage, tons per year__________

Quantity sold as dried pulp, tons per year__________

B. Steffens Waste

Quantity to waste water, untreated, tons per year__________

Quantity to waste water, treated, tons per year________________

Treatment process used________________________

Quantity sold for feed or other end use, tons per year____________

Quantity sold or shipped to other plants for further processing

(for MSG, potash, etc.) tons per year__________________

Quantity otherwise disposed of, tons per year________________

Supplementary Questionnaire on Water Re-Use in 1962

For each recirculation and re-use shown in Section D-II:

1. Would this stream be recirculated if the input water were of excellent quality and there were no restrictions on waste discharge?

If yes, why?

If no, how much more does the recirculation cost than if these conditions prevailed? Indicate facility replacement costs and operating costs separately, net of any incidental reduction in costs due to recirculation.

2. If the recirculation system involves treatment, what happens to the materials removed by the treatment? If these materials are disposed of other than by discharge to a water course or body of water, what is the means and what are the capital and operating costs associated with it?

3. Does the recirculation reduce waste production and its disposal? If so, how and how much? For example, how many more pounds of BOD would be produced per day if the recirculation were not carried out?

INDUSTRIAL WATER USE QUESTIONNAIRE—INSTRUCTIONS

Section A. *General Information*

Data for "typical" periods should be approximately the long-term average.

If 1962 data are not available, report all data in the questionnaire for the latest year available and state the year. (See possible exception under Section E.)

Section B. *Water Use*

1. Fresh water intake (1000 gallons/day). Total water taken from fresh water sources in 1962. This does not include water recycled from one process to another. For comparative purposes, the fresh water intake for 1949 or a year prior to 1962 is needed. This will show changes in water use for industrial purposes.
2. Amount of fresh water taken from specific sources. Show the gallons/day (in 1000's) taken from the specific sources listed. Example: a company using 200,000 gallons/day in 1962 would enter 200 under B-1 above. If half of this fresh water came from wells owned by the plant and half came from a municipal supply which had been derived from a stream, enter 100 opposite plant owned wells and 100 opposite municipal supply (River or Lake).
3. Brackish or salt water used: If plant utilized any brackish or salt water for industrial operations, show average amount per day.
4. Many plants recycle water from one process to another, or re-use water over and over again thereby reducing the total amount of water that otherwise would have been taken from sources listed under 2 and 3 above. If plant had not employed some form of re-use or recirculation, how much additional raw water would have to be obtained from the sources listed under 2 and 3 above?

Section C. *Water and Waste Treatment*

1. Is water treated in plant before use? If the quality of the raw water taken into a plant must be modified in any manner by treatment before it is suitable for use in specific industrial processes, indicate this fact under C-1. In many plants only a portion of the raw water is treated. Please indicate the percentage of total raw water that is treated before use. Do not include recycled or re-used water under this part.

Check the type or types of treatment utilized.

Estimate the present replacement cost of water treatment facilities used to treat the raw water specified above, and estimate annual treatment costs excluding capital costs.

2. Waste water treatment: Include all liquid and liquid-borne wastes discharged from plant. Do not include specific treatment used for preparing waste water for re-use within plant.

Since not all wastes are treated before final discharge, show percentage of total that receive some type of treatment.

Indicate the type of treatment used for treating wastes before they are discharged from plant. Do not include treatment used for preparing wastes for re-use within plant.

Estimate the annual cost of providing the waste treatment specified above. Include overhead, amortization, labor, materials, supplies, etc.

Estimate the total present replacement cost of waste treatment facilities used for treating wastes. Do not include facilities used for treating waste water for re-use within the plant.

Section D-I. *Discharge*

1. Approximate average final effluent discharge from plant (1000 gallons/day). This refers to all water used within plant and finally discarded. In many plants, the final waste discharge volume is approximately equal to the raw water taken into plant. In others, appreciable quantities of water may be lost through evaporation, etc. Include all plant effluents through out-fall sewers except storm water run-off.

2. Approximate daily discharge of treated and untreated process waters to various points. Of the total effluents discarded and not re-used, indicate the amount going to specific points of final disposal. If a portion of a particular process effluent is recycled to the same process or to another process and the balance is discarded, show as discharge only that portion not recirculated.

In many cases, an industrial waste will be stored in an earthen basin for a period of time and later released to a body of water; in such a case the earthen basin would be listed under the forms of treatment, Section C-2, and the final discharge would be listed under Section D-I,2. If the basin has no outlet, list only under Section D-1,2. Show in Section H, Comments, approximate quantity change, if any, of wastes being temporarily retained in a basin (percentage decrease in volume, inlet to outlet) and the approximate quality change (percentage decrease in BOD, total solids, suspended solids).

Section D-II. *Recirculation and Re-Use*

For each process waste water in left column, treated and untreated, which is reused in any way, show quantities reused for each purpose indicated in the five Reuse columns. The last column is for miscellaneous *reuse,* and should not include discharge to sewer or other ultimate disposal. If some sources of water, such as lime cake slurry and Steffens waste, are not reused for *any* purpose, the Reuse spaces on those lines should be left blank. The second column in the table should show the total quantities of each water source, treated and untreated, including all amounts to the various reuses and also all amounts directly to discharge.

Section E. *Raw Water and Final Effluent Quality*

1. Under this section, show specific characteristics of the raw water brought into the plant and of the waste water from the plant. Where several sources of raw water are used, show weighted average values for the total flow.

To get weighted values:

$$\text{ppm} = \frac{\text{Flow } A \times \text{ppm } A + \text{Flow } B \times \text{ppm } B}{\text{Flow } A + \text{Flow } B}$$

Some analyses are not used for both water and wastes. Please report average annual values.

It is understood that some or all types of waste waters may be mixed before treatment or after treatment. In either case, show the quality measurements on the individual stream, if possible. If two or three streams are mixed before treatment, the quality factors before treatment would be different for each stream, and they would all have the same quality factors after treatment. If they are mixed after treatment, qualities of streams would differ from each other both before and after treatment. If one or more wastes receives no treatment, quality "after treatment" should be indicated by "no treatment" or "n.t."

Show data for 1962 if available. If not available, show 1963 data, or state year to which other reported data apply.

1a. Oxygen Consuming Potential: Different industries may use different parameters to measure this quality characteristic. Show values only for those tests which are performed frequently enough to give a realistic average for the raw water and for the waste.

To convert ppm to lb/day:

$$\text{lb/day} = \frac{\text{ppm} \times 8.34 \times \text{daily flow in gallons}}{1{,}000{,}000}$$

Section F. *Water Plans*

1. If plans have been made to change the water treatment or waste treatment system of facilities in any way, please give whatever details are readily available on the changes and estimated cost.

Section G. *Water Ratio*

1. The amount of raw water used per year divided by the gross production per year gives the amount of water used per unit of finished product. Exclude water used solely for power generation. (Give average figures only.)
2. List the type or types of products and the usual units of measure for the items.

Section H. *Comments*

Use this space or attach sheets for answers needing to be qualified or explained. Indicate the section to which the comments apply.

In each instance of a water or waste treatment process or a water recirculation process having been established after the plant was built, show the year in which the particular process was put into operation.

Appendix C

WATER USE SURVEY

SUPPLEMENTARY QUESTIONNAIRE

Company ______________________________ Plant __________

A. Beet tonnage and fresh water withdrawal:

	Now	1949
Typical 24-hr. beet tonnage processed		
Gross fresh water intake, M gal/day		

B. If one or more are used, in what year were the following water and material recovery processes commenced?
Pulp & press water recirculation __________
Cooling water recirculation to flume and beet washers _______
Cooling water recirculation for additional cooling __________
Continuous diffuser __________
Pulp drying __________

C. What changes have occurred in the handling of Steffens waste since the plant was built and when were they introduced?

____

____

____

____

D. If one or more are used, in what year were the following waste processes commenced?
screening ____________ stabilization ponds ____________
sedimentation __________ regulated discharge __________
biological filtration __________ (other specify) __________
evaporation ______________________

E. If a waste pond is used, please furnish the following information:

1. What types of wastes are ponded?

2. If separate ponds are used for different types of wastes, indicate the manner of division.

3. Is a waste pond also used as supply or recirculated water? ________ If so, which pond (if there is more than one) ________________ and what is the use of the recirculated stream? ________________________________

4. What are the total daily and annual volumes of waste (in thousands of gallons) flowing *into* the ponds?
______________ M gal/day ______________ M gal/year

5. What are the total daily and annual volumes of waste (in thousands of gallons) passing *from* the ponds into a stream or lake? (Difference between these figures and those in question 4 is the sum of evaporation and seepage.)
______________ M gal/day ______________ M gal/year
Note: If pond is used for waste flow regulation, or long retention, indicate typical dates of discharge period.
From ________ to ________

6. What is the total pond volume when filled to capacity? ____________

7. What is the approximate average waste retention time in ponds? ____________ days

8. How many pounds of BOD and suspended solids are removed by ponds, per campaign day, from waste waters prior to final discharge?
________ lbs BOD/day ________ lbs suspended solids/day
Note: If discharge from retention ponds is on a different schedule than campaign, compute annual BOD and solids removal and divide by typical number of campaign days.

F. How much fuel is used in the plant per day for all purposes, expressed as tons of coal, gallons of fuel oil, cu. ft. of natural gas, or pounds of steam?

G. Comments: